Path Analysis: Data Analysis Application
Second Edition

Jonathan Sarwono

JONATHAN SARWONO

Copyright © 2018 Jonathan Sarwono
All rights reserved.
ISBN: 9781791743840
ISBN-13: 9781791743840

Published by: Amazon.com, Inc. 410 Terry Avenue North Seattle, Washington 98109 US

DEDICATION

This book is dedicated to Regina Tiatira and Chloe Andrea.

CONTENTS

	Acknowledgments	i
1	Path Analysis Definitions and Basic Concepts	1
2	Assumptions and Basic Principles	11
3	Requirements and Stages in Using Path Analysis	16
4	Path Analysis Models	20
5	An Application of a Multiple Regression Model	25
6	An Application of a Mediation Model	45
7	An Application of a Joint Multiple Regression and Mediation Model	70
8	An Application of a Complex Model	97
9	Path Analysis Using Panel Data	132
10	An Application in Thesis Research	146
11	Indirect Effect Calculation	162

ACKNOWLEDGMENTS

Path Analysis is a popular analysis procedure so far, especially in the field of researches using a quantitative approach. Nevertheless using this procedure is not as easy as we use other analysis procedures, such as correlation and regression. This book is aimed at making easier to use the path analysis by proposing some models and ways how to calculate parameters we are going to find out using IBM SPSS and STATA.

In this opportunity I would like to say thank you to all my books' readers who always welcome to every new book from me. I also thank for my friends' kindness to me, some of them are Prof.Dr. Umi Narimawati, S.E, M.S.i.; Dr Dadang Munandar, M.Si.; Dr. Dewi Indriani Jusuf, S.E, Msi at Computer University of Indonesia and International Women Univeristy, Bandung; Dr. Sawiji at Tarumanagara University Jakarta; Dr. Aristarchus Sukarto at Kridawacana Christian University Jakarta; Dr. Prihartono at PIKSI Ganesha Bandung; Dr. Eddy Suryanto Soegoto, M.Sc. at Computer University of Indonesia.

Those who want to contact me, please visit my web site at http://www.jonathansarwono.info.

Bandung November 2018
Jonathan Sarwono

CHAPTER 1

PATH ANALYSIS DEFINITIONS AND BASIC CONCEPTS

1.1 Definition

What is path analysis? There are many definitions proposed by several experts, some of them are as follows: "Path analysis is an extension of multiple linear regression, and which allows the analysis of more complex models" (Streiner, 2005). Another definition says: "Path analysis is a technique to analyze causal relationships that occur in multiple regression when the independent variables affect the dependent variable not only directly but also indirectly". (Robert D. Retherford 1993). The next definition according to Paul Webley (1997) states: "Path analysis is the direct development of multiple regression forms with the aim of providing an estimate of the magnitude and significance of hypothetical causal relationships in a set of variables." While David Garson (2003) from North Carolina State University defines path analysis as "The regression expansion model used to test the correlation matrix alignment with two or more causal relationships models compared to the researchers. The model is depicted in the form of circle and arrow images where a single arrow shows as a cause. Regression is imposed on each variable in a model as dependent variable while the other as the cause. Regression weighting is predicted in a model compared to the observed correlation matrix for all variables and also the calculation of statistical goodness of test. While according to the author, path analysis is an analytical technique used to analyze the inherent causal relationships between variables that are arranged based on the temporary sequence by using path coefficient as the amount of value in determining the magnitude of the effect of exogenous independent variables on the endogenous dependent variable (Sarwono, 2011). From the above definitions, it can be concluded that the path analysis can be said as an extension of multiple regression analysis. Though based on history there is a basic difference between path analysis that is independent of statistical procedures in determining the cause and effect relationship; while linear regression is indeed a statistical procedure used to analyze the causal relationships among the variables studied.

1.2 Objectives to Use Path Analysis

The objectives of using path analysis are: a) Viewing relationships between variables based on a priori model; b) Explaining why variables correlate each other by using a sequential temporary model; c) Drawing and testing a mathematical model using the underlying equation; d) Identifying the causal path of a particular variable affecting other variables; e) Calculating the strength of the effect of one or more exogenous variables on other endogenous variables.

1.3 Advantages and Disadvantages of Path Analysis

Advantages of using path analysis include: a) Ability to test the overall model and individual parameters; b) Modeling ability of several mediator and intervening variables; c) The ability to estimate by using equations that can see all possible causal relationships across all the variables in the model; d) The ability to decompose correlations into causal relationships, such as direct effects and indirect effects and non-causal associations, such as spurious components. While disadvantages of using path analysis include: a) inability to reduce the impact of measurement error; b) Path analysis only has the variables that can be observed directly; c) Path analysis does not have indicators of a latent variable; d) Since path analysis is an extension of multiple linear regression, then all assumptions in this formula must be followed; e) The causes in the model are only one direction (specifically called as *recursive*) ; should not be reciprocal

1.4 Basic Terms

Before studying more deeply what path analysis is; it will be explained some basic important terms that will help in understanding the path analysis correctly by using the following diagrams.

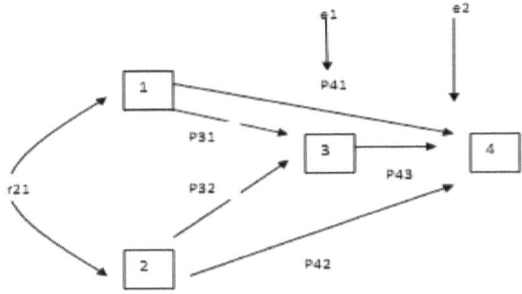

Figure 1.1 Path Diagram Model

Path model. The path model is a diagram that links the network of relationships of several variables placed in sequence to be studied in the research. The conventional term is the association between independent, intervening and dependent variables. In the following section we will discuss this more deeply because in the path analysis it is actually not known the terms of independent and dependent variables. The pattern of relationships in path analysis is indicated by using arrows. The single arrows indicate a causal relationship between the independent variables which in the path analysis is then referred to as exogenous variables (1 and 2) variables with one or more dependent variables which is in the path analysis referred to as the endogenous variable (3 and 4). The 3 variable funcions as an intervening variable. The arrows also connect errors (residue variables), e1 and e2 with all endogenous variables respectively. Double arrows show the correlation between pairs of exogenous variables (r21). In the picture above, there are two parts, namely: 1 and 2 to 3 variables called as sub-structure I and from 1,2 and 3 to 4 called sub-structure II. **The causal path for the existing variables** include the first directional paths of the arrows leading to the variable coming from the previous variable, e.g. direct arrows from 1 to 3, 2 to 3 and from 3 to 4 variables as well from 1 and 2 to 4 which can be seen the following figure using the bold arrows.

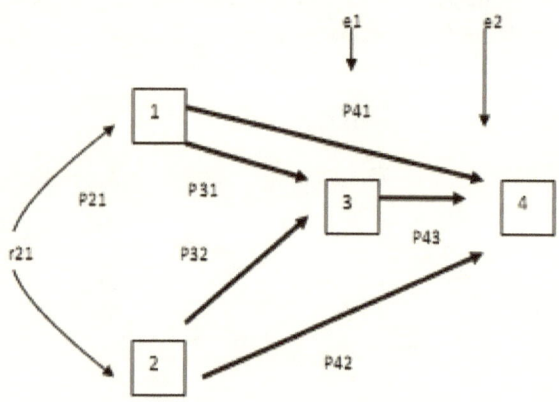

Figure 1.2 Causal Path 1 and 2 to 3 and to 4 (sub structure I) 3 to 4 (sub structure II)

Exogenous Variables. Exogenous variables in a path model are all variables with no explicit causes or in the diagram there are no arrows leading in, other than the measurement error section. This variable serving as the independent variable causes the next sequence variable called the endogenous variable (e.g variables of 1 and 2 towards 3 in sub structure I and 3 to 4 in sub structure II). If between exogenous variables are correlated then the correlation is indicated by a two-way arrow connecting those variables (1 is correlated with 2). For details see the following picture in the bold box section.

PATH ANALYSIS: DATA ANALYSIS APPLICATION

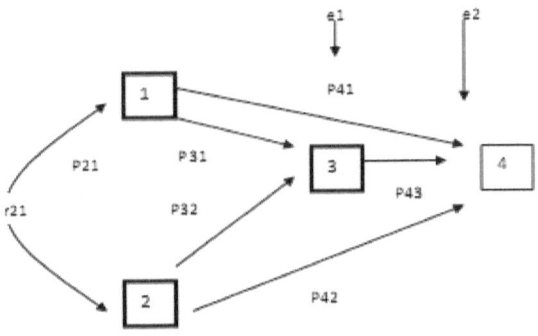

Figure 1.3 Exogenous Variables of 1 and 2 (sub structure I) and 3 (sub structure II)

Endogenous Variables. Endogenous variables are variables that have arrows towards the variable (variables of 3 for sub-structures I and 4 for sub-structures 2). The variables include all intervening and dependent variables. An endogenous intervening variable has arrows leading into it and at the same time coming from the direction of the variable in a path diagram model. While a dependent variable only has arrows that lead to it. For details see the following picture in the bold box section

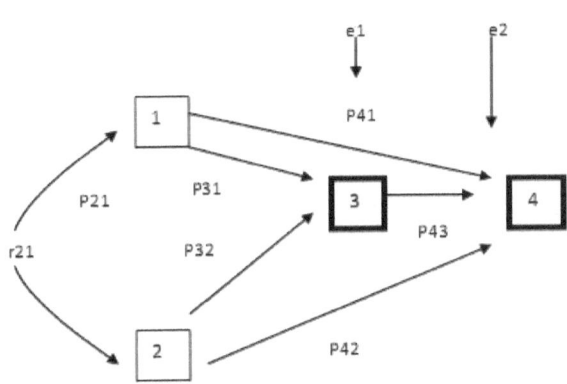

Figure 1.4 Endogenous Variable 3 (sub structure I) and 4 (sub structure II)

Path Coefficient / Path Weighting (p). Path coefficient is a standardized coefficient called 'beta weight' which indicates the direct influence of an exogenous variable on an endogenous variable in a particular path model (p31 and p32 for sub-structure I and p41, p42, and p43 for sub-structure II). Therefore, if a model has two or more causal variables, the path coefficients are partial regression coefficients that measure the magnitude of the influence of one variable on another variable in a particular path model that controls the other two previous variables using data which have been standardized or use a correlation matrix as the input. The standardized path coefficient has a coefficient meaning that has been converted to a standard Z value that allows the researcher to compare the relative strength of the effects of all the different exogenous variables in a particular path model. For details see the following picture in the bold one-way arrow.

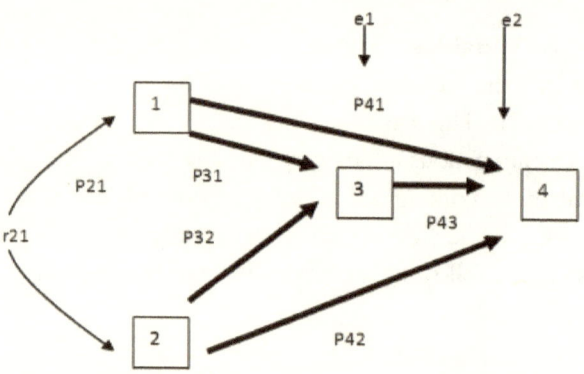

Figure 1.5 Path Coefficient p31 and p32 (sub structure I) and p41, p42, and p43 (sub structure II)

Correlated Exogenous Variables. If all exogenous variables are correlated, then as a marker of the correlation is a two-way arrow connected between the variables with their correlation coefficients (variables of 1 and 2 with the correlation called r21). For details see the following figure in the bold two-way arrows.

PATH ANALYSIS: DATA ANALYSIS APPLICATION

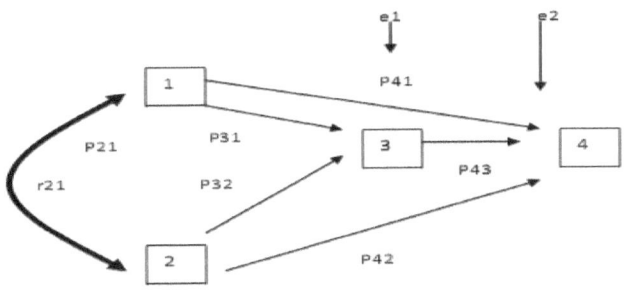

Figure 1.6 Exogenous Variables of 1 and 2 correlated (r_{21})

Error / Error Term (e). The term residual error, technically referred to as 'residual' or 'residue' reflects the unexplained variant or the influence of all the non-measurable exogenous variables directly plus the measurement errors (e1 and e2) which reflect the cause of the unknown variability in the analysis results. The magnitude of the effect of this error on endogenous variables that reflect other unmeasured variables is $1 - r^2$. The magnitude of the value of the variant is $1 - r^2$ multiplied by the variant of the value of the endogenous variable. Residual errors are assumed to be normally distributed and have mean of 0 and do not correlate with all the variables in the model under investigation. For details see the diagram below in the bold arrows.

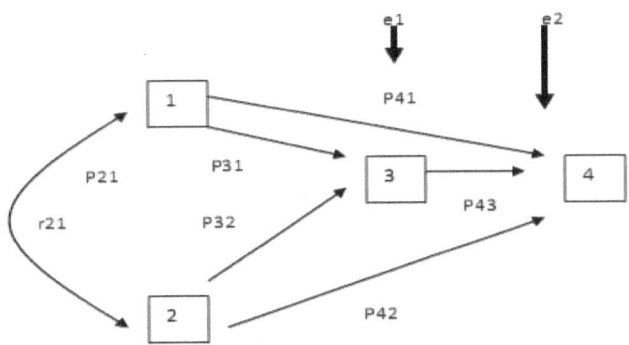

Figure 1.7 Error Term of e1 for sub structure I and e2 for sub structure II

Path Multiplication Rules. The value of a composite path is the result of all path coefficients or all path coefficients of all exogenous and endogenous variables.

Effect Decomposition. Path coefficients can be used to parse correlations in a model into direct and indirect effects associated with direct and indirect paths reflected by arrows in a particular model. It is based on the rule that in a linear system, the total cause effect of an 'i' variable to the variable 'j' is the sum of all path values from "i" to "j".

Correlation decomposition. the correlation between two variables can be composed into four components as follows: a) Direct effect of exogenous variables (x) to endogenous ones (y); b) The indirect effect of exogenous variables to endogenous ones through an intervening variable; c) Components that we can not analyze because of the lack of our knowledge of the causal direction in a single path; d) Spurious components caused by exogenous (x) and endogenous (y) variables that are respectively affected by a particular third variable or set of other variables in the model.

Significance and Goodness of Fit in a Model. To perform the individual testing of path coefficients, we can use standard t values or F tests of regression. To test the model with all the paths, we can use the goodness of fit test using the significance (probability) value (sig / p value) in the ANOVA section. If a model is correct, it includes all appropriate variables and excludes all non-conforming variables; then the number of path values from i to j will be equal to the regression coefficient for j predicted based on i, that is for the standardized data in which simple regression coefficients are equal to the correlation coefficients; then the sum of all coefficients (standards) will be equal to the correlation coefficient.

One-way and Two-way Arrows. If you want to describe the cause, then use arrows in one direction. To illustrate the correlation, use a curved arrow in both directions. There are times when a causal relationship produces a negative number, to illustrate the negative results used dashed lines.

PATH ANALYSIS: DATA ANALYSIS APPLICATION

Pattern of Relationships. In the path analysis, we do not use the terms of independent or dependent variables. Instead we use exogenous variables for the independent ones and endogenous variables for the dependent ones.

Recursive Models. The one-way causal model. There is no reverse direction (feedback loop) and no reciprocal effect. In this model one variable cannot function as a cause and effect at the same time. Thus in path analysis, it is known as a recursive model only.

Non-Recursive Models. The causal model accompanied by a feedback loop or a reciprocal effect. This relationship model is not known in Path Analysis and is known only in Structural Equation Modeling (SEM).

Direct Effect (DE). The direct effect can be seen from the path coefficient of one exogenous variable to an endogenous one. In the Figure 1.1 example, 1 and 2 to 3 as well as 1 and 2 to 4

Indirect Effect (IE). The sequence of paths through one or more intervening variables. To get the value, it can done by multiplying the path coefficient from the exogenous variable to the intervening one with the path coefficient from that intervening variable to the endogenous variable. In the Figure 1.1 example, p_{31} times p_{43} and p_{32} times p_{43}

Total Effect (TE). The sequence of paths through one exogenous variable to the intervening one plus from that intervening variable to the endogenous one. To get the value, it can be done by adding the path coefficient from the exogenous variable to the intervening one with the path coefficient from that intervening variable to the endogenous one. In the Figure 1.1 example, p_{31} plus p_{43} and p_{32} plus p_{43}

Combined Effect (R^2): The combined effect is the effect of all exogenous variables on the endogenous variables whose value is called as R^2 which is also used to assess the goodness of fit of the research model with theoretical one.

Partial Effect (P): The partial effect is the one-by-one effect of each exogenous variable on the endogenous variable whose value is termed as the path coefficient / beta value. In the Figure 1.1 example, p31, p32, p41, p42 and p43

Influence of Other Factors (Error): The influence of other factors is the influence of other variables beyond the path model studied. The value is obtained by the formula: $e = 1 - r^2$.

CHAPTER 2

ASSUMPTIONS AND BASIC PRINCIPLES

When we want to use the Path Analysis procedure, here are some of the basic assumptions and principles:

Linearity. Relationship between variables are linear. It means that the data distribution forms a straight line from bottom left to top right, as shown below:

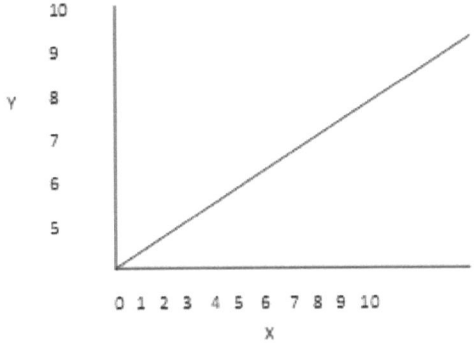

Figure 2.1 Linearity

Co-linear. Showing a similar line. It means, if there are some exogenous variables affecting one endogenous variable; or otherwise an exogenous variable affects some endogenous variables if it is drawn a straight line will form the same lines.

The Chain Model of Effect: Indicating a causal model in which the sequence of events leads to variation in the endogenous variable, as shown below. In the picture below all sequences of events X1, X2, X3, and X4 go to Y

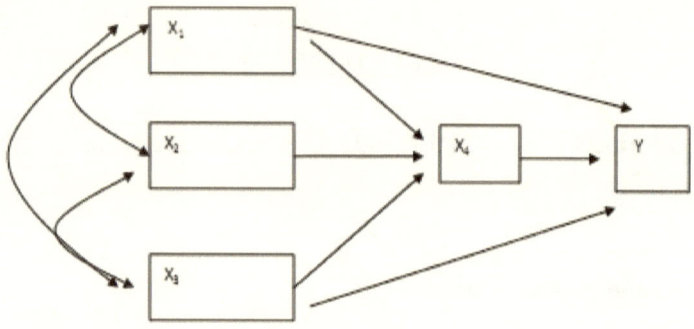

Figure 2.2 Causal Relationship Model

Aditivity. Aditivity means that there are no interaction effects

Causal closure: All direct effects of one variable on another variable must be included in the path diagram.

Coefficient of Beta (β). It is a standardized regression coefficient showing the number of changes in the endogenous variable associated with the unit change (increase or decrease) in a standard deviation on the exogenous variable when controlling the effect on other independent variables. Coefficient of beta is also called beta weight (β). This value is used as a value quantity in the path coefficient (p) or the amount of influence of each exogenous variable on the endogenous variable individually or referred to as a partial effect.

Coefficient of Determination (R^2): Also referred to as the association index. It is a value that shows how much variation in one variable is defined or explained by one or more other variables and how variation in one variable are related to the variation in other variables. In bivariate statistics it is abbreviated as r^2 and in multivariate is abbreviated as R^2. This value is used as a value scale to express the magnitude of the effect of all exogenous variables on endogenous variables simultaneously or referred to as the combined effect.

Metric data. All observed variables have scaled values. If the data has not been in the form of interval scale, the data should be modified using Method of Successive Interval / MSI first. If non metric data is used then it will shrink the correlation coefficient value. Small correlation coefficient value will cause the value of R^2 becomes smallest. Accordingly, the modeling created using the path analysis will not be valid; because one indicator of the goodness of fit in model pursuant to the theory is to see the value of R^2 approaching 1. If this value is closer to 1; then the model is considered good or in accordance with the theory.

Residual variables are not correlated with any of the variables in the model.

Disturbance terms or residual variables should not be correlated with all endogenous variables in the model. If violated, it will result in incorrect regression results to estimate path parameters.

Low multicolliniearity. Multicollinearity means two or more independent variables (exogenous variables) have very high correlation. If there is high correlation then we will get a large standard error of beta coefficient (B) used to eliminate ordinary variance in conducting partial correlation analysis.

Recursivity. All arrows have one direction, no looping or no reciprocal relationship

The correct model specification is required to interpret path coefficients. A specific error occurs when a significant cause variable (an exogenous one) is removed from the model. All path coefficients will reflect covariance along with all unmeasured variables and will not be interpreted appropriately in relation to direct and indirect effects.

Correct input. This means that if we use the correlation matrix as the input, the Pearson correlation is used for two interval-scale variables; polychoric correlation for two ordinal ordinal variables; tetrachoric for two dichotomous variables (nominal scale); polyserial for one interval and other ordinal variables; and biserial for one variable of interval and other nominal scale.

An adequate sample size. Use a sample of at least 100 with a 10% error rate to obtain a significant and more accurate analysis result. For ideally large samples of 400 - 1000 are generally requirements in multivariate analysis techniques (using error rate of 5%).

Design the model according to existing theory to show a causal relationship in the variables being studied. The causal relationship is inherent to the theory and independent to statistics. That is why the role of theory is critical when we use the path analysis.

Because of the calculation of path analysis uses linear regression techniques; then the general assumption of linear regression should be followed, namely: 1. The regression model should be feasible. This feasibility is known if the significance level in ANOVA is <0.05; 2. . Predictor used as a free variable should be feasible. This feasibility is known if the Standard Error of Estimate <Standard Deviation. 3. . Regression coefficients must be significant. The test is done by t test. Regression coefficient is significant if t count > t table (critical value). 4. There should be no multicollinearity. 5.

No autocorrelation occurs. Autocorrelation does not occur if the value of Durbin and Watson is $-2 \leq DW \leq 2$

CHAPTER 3

REQUIREMENTS AND STAGES IN USING PATH ANALYSIS

3.1 Requirements

The absolute requirement that must be met when we will use path analysis in addition to what has been discussed in detail in previous chapter, it is recommended that some of these requirements should not be violated: a. Use metric data instead of non-metric one; b. Exogenous and endogenous variables exist for multiple regression models and intervening variables for mediation models and combined models of mediation and multiple regression as well as complex models; e. Sufficient sample size is preferably as much as 100 and ideally 400 – 1000; f. Pattern of relationship between variables is only one direction, therefore there should be no reciprocal relationship; g. Causal relationships are based on existing theory with the previous assumption that there is a causal relationship in the variables we are examining.

3.2 Stages in Using Path Analysis

Stages in employing path analysis are:

1. **Designing models based on theory.** For example we will see the effect of product quality, price and service quality variables to the level of customer satisfaction. Departing from the existing theory then we make a hypothesized model.

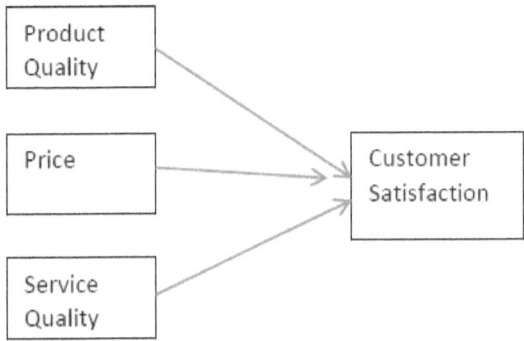

Figure 3.1 A Model of Product Quality, Price, Service Quality and Customer Satisfaction Relationship Based on Theory

2. **The model hypothesized**: In this section we create a hypothesis that states, for example:

H0: Variables of product quality, price and service quality do not affect the level of customer satisfaction either simultaneously or partially.

H1: Variables of product quality, price and service quality affect the level of customer satisfaction either simultaneously or partially

3. **Path diagram model**. Determine the path diagram model based on the variables studied.

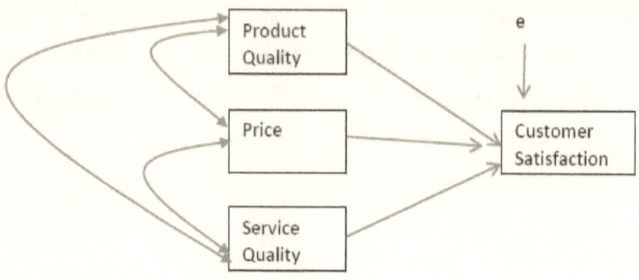

Figure 3.2 Path Diagram Model

4. **Path diagram** . Finally we create a path diagram as follows:

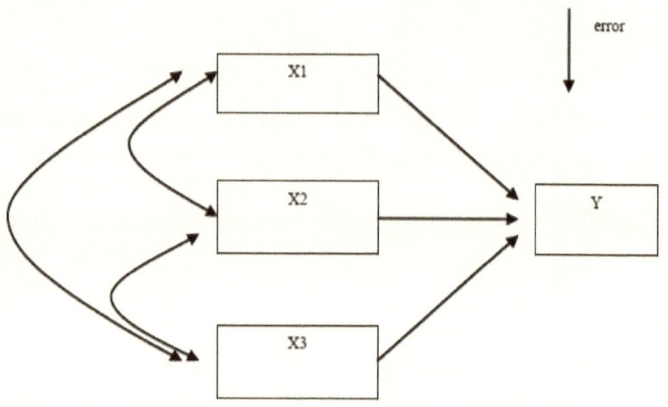

Figure 3.3 Path Diagram

Where:
X1 as an exogenous variable of product quality
X2 as an exogenous variable of price
X3 as an exogenous variable of service quality
Y as an endogenous variable of customer satisfaction

5. **A Structural equation**. Creating a structural equation as follows The path diagram above the structural equation is: Y = PYX1 + PYX2 + PYX3 + e

6. **Perform path analysis procedures with SPSS.** To calculate tha data we will use the SPSS software

7. **Calculating values of parameters.** Some values that will be calculated are: combined effect, partial effect, direct effect, indirect effect, total effect, effect of other factors, correlation, and validity test

CHAPTER 4

PATH ANALYSIS MODELS

There are four models in path analysis. The four models will discussed one by one in the next session: a multiple linear regression model, a mediation model, a joint (combined) model of multiple regression and mediation, and a complex model.

4.1 A Multiple Linear Regression Model
This multiple regression model is actually the development of multiple linear regression analysis using more than one exogenous variables and one endogenous variable. In this example below, the exogenous variables are X1 and X2 and the endogenous variable is Y. This model has a path diagram as shown below:

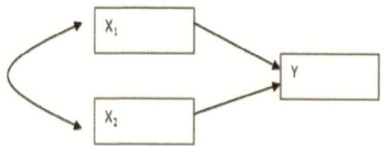

Figure 4.1 A Multiple Linear Regression Path Analysis Model

Where:
- X1 is the first exogenous variable
- X2 is the second exogenous variable
- Y is an endogenous variable

PATH ANALYSIS: DATA ANALYSIS APPLICATION

Sample of the case: a researcher wants to see the magnitude of the effect of product quality and advertising variables on sales, then X1 is the product quality variable and X2 is the advertising variable whereas Y is the sales variable. In the path analysis terminology, product quality and advertising are exogenous variables and sales is an endogenous variable.

4.2 A Mediation Model

The second model is the mediation model or intervening variable one where the presence of Y variable as an intervening variable will change the effect of variable X to variable Z. This influence can be a decrease or an increase. The second model has the path diagram as shown below:

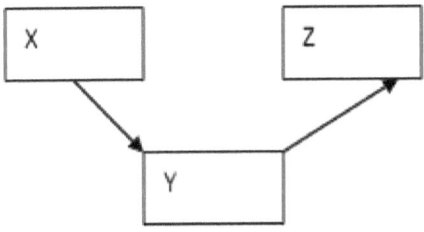

Figure 4.2 A Path Analysis Mediation Model

Where:
- X is an exogenous variable
- Y is an intervening endogenous variable
- Z is an endogenous variable

Sample of the case: In order to increase profits, a company sells products at a low price by ignoring the quality of the product itself. The result does not increase the profit; nevertheless it even decreases. If applied in this second model, then X is product, Y is product quality variable and Z is a profit variable. Thus, the product variable affects the profit variable through the product quality variable

4.3 A Joint (Combined) Model between Multiple Regression and Mediation Models

The third model is the combination between a multiple linear regression model with a mediation one, that is X variable affects Z variable directly (direct effect) and indirectly (indirect effect) also affects Z variable through intervening Y variable.

In this model it can be explained as follows: The X variable serves as an exogenous variable to the Y and Z endogenous variables. Y variable has two functions: the first function is as an endogenous variable to the X exogenous variable; the second function is as an intervening endogenous variable in order to see the effect of X on Z through Y. The Z variable is an endogenous variable.

This model has a path diagram as below:

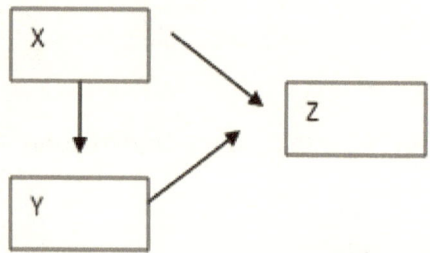

Figure 4.3 A Joint (Combined) Path Analysis Linear Regression and Mediation Model

Where:
- X is an exogenous variable
- Y is an endogenous variable whose function is as an intervening variable
- Z is an endogenous variable

Sample of the case: In this case X variable is service quality, Y variable is customer satisfaction and Z variable Z is customer loyalty.

4.4 A Complex Model

The fourth model is a complex one, namely the X1 variable directly affects Y2 and through the X2 variable, it also indirectly affects Y2, while Y2 is also affected by Y1 variable. In this model it can be explained as follows: The X1 variable serves as an exogenous variable. X2 variable has two functions, namely the first function is as an endogenous variable to the X1 exogenous variable; the second function is as an intervening endogenous variable in order to see the effect of X1 on Y2 through X2. The Y2 variable is an endogenous variable while the Y1 is an exogenous variable. This model has a path diagram as shown below.

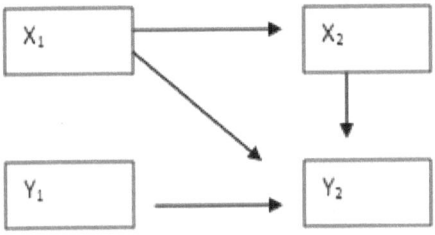

Figure 4.4 A Complex Model

Where:
o X1 is an exogenous variable
o X2 is an endogenous variable whose function is as an intervening variable
• Y1 is an exogenous variable
• Y2 is an endogenous variable

Sample of the case: For the above model we can make an example as follows: X1 is an employee performance variable; X2 is a service quality variable; Y2 is a customer satisfaction variable; and Y1 is the product quality variable. In one company the performance of employees will affect the quality of service directly and indirectly will affect the customer satisfaction through service quality. And product quality will also affect the level of customer satisfaction.

JONATHAN SARWONO

CHAPTER 5

AN APPLICATION OF A MULTIPLE REGRESSION MODEL

5.1 A Case Example

In this example we will use three independent variables that serve as exogenous variables and one dependent variable that serves as an endogenous variable. As exogenous variables are product quality, service quality and complaint handling variables, while as an endogenous variable is the level of customer satisfaction. The relationship between these variables when depicted in the path diagram model becomes as follows:

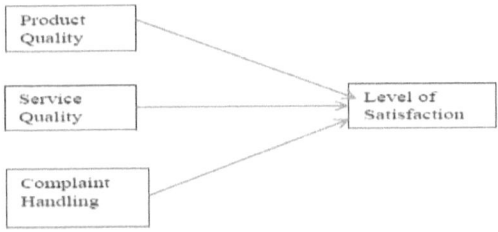

Figure 5.1 A Path Diagram Model on the Effect of Product Quality, Service Quality and Handling Complaint Variables on Customer Satisfaction Level Variable Based on Theory

5.2 Hypotheses and Problem Formulations

The hypothesis for this case is:
H0: Product quality, service quality and complaint handling variables have no effect on customer satisfaction level either simultaneously or partially
H1: Product quality, service quality and complaint handling variables have effect on customer satisfaction level either simultaneously or partially

The problem formulations for this case are:
a. How much is the effect of product quality, service quality and complaint handling on customer satisfaction partially?
b. How much is the effect of product quality, service quality and complaint handling on customer satisfaction simultaneously?

5.3 Research Data

There are 100 data for the above case as presented in the table below:

No	product	service	complaint	satisfaction	No	product	service	complaint	satisfaction
1	18	17	16	15	51	13	14	13	12
2	15	17	18	12	52	13	13	14	10
3	17	14	16	14	53	15	16	16	15
4	14	14	14	13	54	12	12	14	12
5	15	15	16	12	55	13	12	15	9
6	17	15	16	13	56	15	15	16	14
7	13	16	12	14	57	14	18	12	10
8	19	19	20	11	58	13	11	13	11
9	15	16	17	14	59	10	11	13	12
10	19	19	18	14	60	16	18	16	12
11	14	16	17	12	61	14	11	11	10
12	15	11	15	10	62	17	12	12	13
13	14	14	13	12	63	16	14	13	12
14	16	16	18	11	64	14	15	12	10
15	10	14	16	12	65	13	11	14	10
16	13	15	17	13	66	13	11	14	11
17	19	12	17	14	67	12	11	14	9
18	15	12	16	14	68	13	14	13	10
19	15	12	14	15	69	15	12	13	9
20	16	14	15	14	70	13	13	17	12
21	12	11	15	14	71	13	14	12	11
22	18	16	14	14	72	13	16	15	12
23	13	15	15	11	73	15	14	15	12
24	14	14	13	12	74	14	13	15	10
25	13	13	17	11	75	14	14	13	10
26	10	10	12	9	76	14	16	18	15
27	13	17	13	9	77	20	19	22	17
28	12	12	11	10	78	17	20	20	16
29	11	14	12	8	79	16	20	20	15
30	9	14	8	7	80	18	16	16	16
31	10	10	13	12	81	15	18	19	12
32	8	12	12	9	82	16	19	19	12
33	13	14	12	10	83	20	16	15	13
34	8	15	14	9	84	21	22	22	19
35	11	8	12	10	85	19	19	19	12
36	13	13	16	11	86	18	15	16	13
37	11	14	11	9	87	18	19	20	12
38	9	15	14	11	88	17	15	18	12
39	15	11	12	12	89	21	17	20	17
40	11	12	15	8	90	16	17	19	17
41	11	12	13	8	91	16	19	18	16
42	9	14	14	9	92	19	16	20	16
43	10	17	11	9	93	17	19	18	16
44	9	10	12	8	94	19	18	21	13
45	12	9	12	7	95	21	17	20	19
46	13	15	15	8	96	19	16	17	15
47	13	8	14	10	97	19	15	19	15
48	14	11	11	9	98	17	19	19	15
49	11	13	10	9	99	19	19	19	12
50	12	13	15	13	100	18	17	19	14

5.4 Stages in Completing the Case

To solve the case above the stages are as follows:

First: Create a model of the path diagram based on the relationship among variables we are studying

The path diagram model of the relationship among variables we are studying is as follows:

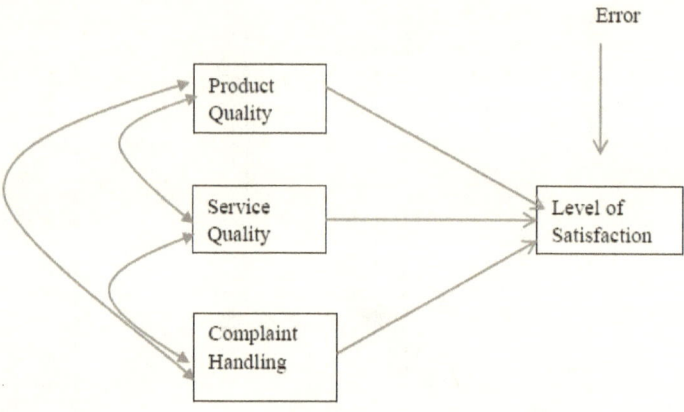

Figure 5.2 A Path Diagram Model on the Effect of Product Quality, Service Quality and Handling Complaint Variables on Customer Satisfaction Level

Second: Create a path diagram of the above model

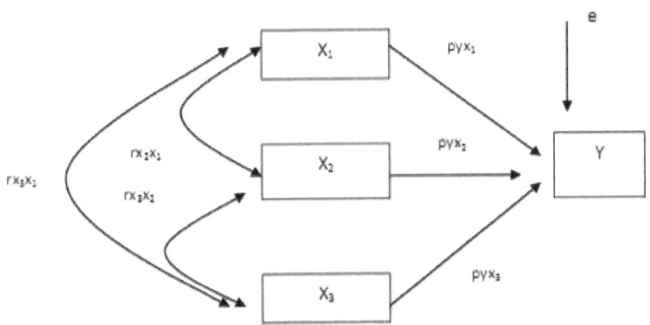

Figure 5.3 A Path Diagram on the Effect of Product Quality, Service Quality and Handling Complaint Variables on Customer Satisfaction Level

Where:
X1 as a product quality exogenous variable
X2 as a service quality exogenous variable
X3 as a complaint handling exogenous variable
Y as a satisfaction level endogenous variable

Third: Create a structural equation
In the path diagram above there is only one structural equation. The structural equation is as follows:

$$Y = PYX_1 + PYX_2 + PYX_3 + e$$

Fourth: Create a variable design, enter data and analyze it in IBM SPSS Design variables can be created by selecting the left bottom sub menu on the command: **Variable View**. Write down all variables pursuant to the example below.

Name	Type	Width	Decimal	Label	Values	Missing	Column	Align	Measure	Role
product	Numeric	8	0	Product Quality	None	None	8	Right	Scale	Input
service	Numeric	8	0	Service Quality	None	None	8	Right	Scale	Input
complaint	Numeric	8	0	Complaint Handling	None	None	8	Right	Scale	Input
satisfaction	Numeric	8	0	Satisfaction level	None	None	8	Right	Scale	Target

Fifth: Enter 100 data by clicking on the command: **Data View**

No	product	service	complaint	satisfaction
1				
.				
.				
100				

Sixth: Perform analysis by using IBM SPSS with the following steps

- Click **Analyze> Linear**
- Enter the satisfaction variable to the **Dependent** column
- Enter the product quality, service quality and complaint handling variables to **Independent** column
- Fill in the **Method** field with the **Enter** command
- Click on **Option**: In the **Stepping Method Criteria** option enter 0.05 in the **Entry> Check Include constant in equation** > In the **Missing Values** option check **Exclude cases listwise** > Press **Continue**
- Select **Statistics**: In the **Regression Coefficient** option select **Estimate, Model Fit, Descriptive and Durbin - Watson**. In the **Residual** option, select **Case wise Diagnostics** and check **All Cases** > Press **Continue**
- Click on **Plots** to create **Graph** > Fill in column Y with SDRESID and X option with ZPRED, then press **Next** > Fill in column Y with ZPRED and X column with DEPENDNT > In **Standardized Residual Plots** option, check **Normal Probability Plot** and **Histogram** > Press **Continue**
- Click **Ok** to process

PATH ANALYSIS: DATA ANALYSIS APPLICATION

The results of the calculation is as follows:

First Part: Output of the Regression Section

Model Summary[b]

Model	R	R Square	Adjusted R Square	Std. Error of the Estimate	Durbin-Watson
1	,773[a]	,597	,584	1,685	1,911

a. Predictors: (Constant), Product Quality, Service Quality, Complaint Handling
b. Dependent Variable: Customer Satisfaction

ANOVA[b]

Model		Sum of Squares	df	Mean Square	F	Sig.
1	Regression	403,413	3	134,471	47,365	,000[a]
	Residual	272,547	96	2,839		
	Total	675,960	99			

a. Predictors: (Constant), Product Quality, Service Quality, Complaint Handling
b. Dependent Variable: Customer Satisfaction

Coefficients[a]

Model		Unstandardized Coefficients		Standardized Coefficients	t	Sig.
		B	Std. Error	Beta		
1	(Constant)	1,327	,968		1,371	,173
	Product Quality	,381	,080	,457	4,758	,000
	Service Quality	,034	,082	,038	,421	,675
	Complaint Handling	,305	,093	,345	3,273	,001

a. Dependent Variable: Customer Satisfaction

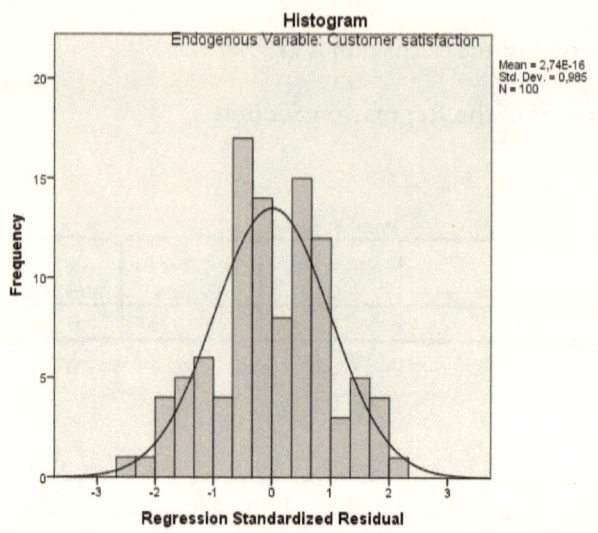
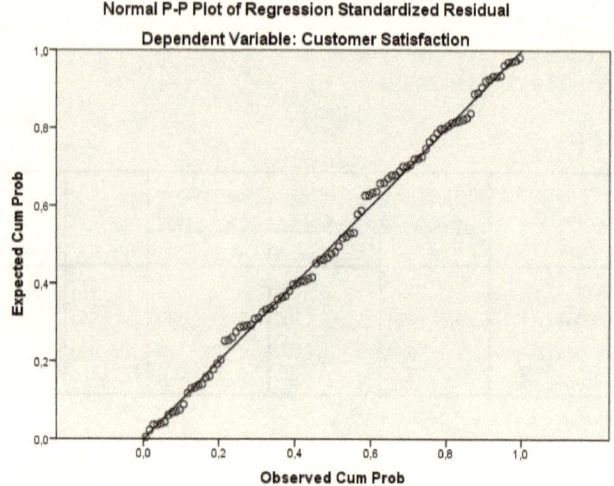

PATH ANALYSIS: DATA ANALYSIS APPLICATION

Second Part: Output of Correlation Section

Correlations

		Customer Satisfaction	Product Quality	Service Quality	Complaint Handling
Pearson Correlation	Customer Satisfaction	1,000	,730	,548	,702
	Product Quality	,730	1,000	,600	,724
	Service Quality	,548	,600	1,000	,684
	Complaint Handling	,702	,724	,684	1,000
Sig. (1-tailed)	Customer Satisfaction	.	,000	,000	,000
	Product Quality	,000	.	,000	,000
	Service Quality	,000	,000	.	,000
	Complaint Handling	,000	,000	,000	.
N	Customer Satisfaction	100	100	100	100
	Product Quality	100	100	100	100
	Service Quality	100	100	100	100
	Complaint Handling	100	100	100	100

5.5 The interpretation of the calculation results

Part One: Multiple Linear Regression Analysis

In this section, the interpretation is divided into two: first, looking at the combined effect of the three exogenous variables on one endogenous variable, and second, looking at the partial effect for each of the exogenous variable on one endogenous variable.

The Effect of Exogenous Product Quality, Service Quality, and Complaint Handling Variables on Customer Satisfaction Level

The effect of product quality, service quality, and complaint handling exogenous variables on the customer satisfaction level variable can be seen in the output in the **Model Summary** table, in the R square value. The value of R square (R^{2}) in the table above is 0.597. This number shows the amount of variation of the customer satisfaction endogenous variable that can be explained by exogenous variables of product quality, service quality, and complaint handling. In another way, we can say the R square (R^{2}) of 0.597 shows the amount of the effect of exogenous variables of product quality, service quality, and complaints handling on the customer satisfaction endogenous variable.

This value can be made in the form of percent numbers by calculating Coefficient of Determination (CD) by using the following formula.

CD = R^2 x 100%

CD = 0.597 x 100%

CD = 59.7%

While the rest can be calculated using the following formula:

e = 1 - R^2

e = 1 − 0.597

e = 0.403

The number 0.403 (40.3%) is the influence of other factors in the model beyond the three exogenous variables studied. In other words, the variation of customer satisfaction level which can be explained by using exogenous variables of product quality, service quality, and complaint handling is equal to 59.7%; while the 40.3% is caused by other variables outside of this research.

Simultaneous Hypothesis Testing to See The Effect of Exogenous Product Quality, Service Quality, and Complaint Handling Variables on a Customer Satisfaction Level Variable

This hypothesis testing is used to see the simultaneous effect of the exogenous product quality, service quality, and combined complaint handling variables on customer satisfaction level. The testing uses the value of significance (sig) from the ANOVA output table. To see whether there is a linear relationship between product quality, service quality, and combined complaint handling with customer satisfaction level, we can perform the following analysis steps:

First: Determine the hypothesis to be tested
H0: There is no linear relationship between product quality, service quality, and complaint handling with customer satisfaction level
H1: There is linear relationship between product quality, service quality, and complaint handling with customer satisfaction level

Second: Calculate the value of significance (sig) or pvalue
The sig value shown in column Sig is 0.000

Third: use the following criteria
If the sig value < 0.05 reject H0 and accept H1
If the sig value > 0.05 accept H0 and reject H1

Fourth: take the decision of the hypothesis testing

Since the sig value is 0.000 < 0.05; therefore reject H0 and accept H1. Thus there is linear relationship between product quality, service quality, and combined complaint handling with customer satisfaction level. Accordingly product quality, service quality, and complaint handling affect customer satisfaction level significantly.

The Partial Effect of Product Quality, Service Quality, and Complaint Handling Exogenous Variables on A Customer Satisfaction Level Endogenous Variable

The magnitude of the effect of exogenous variables of product quality, service quality, and complaint handling individually / partially on a customer satisfaction level endogenous variable can be seen from the Beta or Standardized Coefficients value. Whereas the testing of hypothesis uses t value. The values can be seen in the **Coefficient** table output. Therefore, in the following sections we will discuss the effect of exogenous variables of product quality, service quality, and complaint handling on a customer satisfaction endogenous variable partially.

Relationship between a Product Quality Exogenous Variable and a Customer Satisfaction Level Endogenous Variable

To see whether there is a linear relationship between a product quality exogenous variable and a customer satisfaction endogenous variable we can perform the following analysis steps:

First: Determine the hypothesis to be tested
H0: There is no linear relationship between a product quality exogenous variable and a customer satisfaction endogenous variable
H1: There is linear relationship between a product quality exogenous variable and a customer satisfaction endogenous variable

Second: Calculate the value of observation t (t_o)
The t_o value shown in column t in the above **Coefficients** table is 4.758

Third: Calculate the number of t table (t_α) with the following provisions Determine the level of significance of 0.05 and Degree of Freedom (DF) with the provisions: DF = n -2, or 100 - 2 = 98. From these provisions it is obtained t_α as much as 1.960

Fourth: Define the decision-making criteria as follows
If $t_o > t_\alpha$, then H0 is rejected and H1 accepted;
If $t_o < t_\alpha$, then H0 is accepted and H1 is rejected
or we can use the second alternative by using the significance level (probability / p value)
If sig < 0.05, then there is significant influence
If sig > 0.05, then there is no significant influence

Fifth: Make a decision of hypothesis testing results
The calculation result with IBM SPSS shows the t_o as much as 4.758 is bigger than t_α as much as 1.960; thus the decision is reject H0 and accept H1. This means that there is a linear relationship between a product quality exogenous variable and a customer satisfaction level endogenous variable. Because there is a linear relationship between the two variables; then the product quality affects the customer satisfaction level significantly. The amount of effect can be known from coefficient value of Beta (in column **Standardized Coefficient Beta**) as much as 0.457. The effect of this

magnitude is significant because the value of significance in the **Sig column** is 0.000 which is smaller than 0.05.

Relationship between a Service Quality Exogenous Variable and a Customer Satisfaction Level Endogenous Variable

To see whether there is a linear relationship between a service quality exogenous variable and a customer satisfaction endogenous variable we can perform the following analysis steps:

First: Determine the hypothesis to be tested
H0: There is no linear relationship between a service quality exogenous variable and a customer satisfaction endogenous variable
H1: There is linear relationship between a service quality exogenous variable and a customer satisfaction endogenous variable

Second: Calculate the value of observation t (t_o)
The t_o value shown in column t in the above **Coefficients** table is 0.421

Third: Calculate the number of t table (t_α) as it has been discussed above
From these provisions it is obtained t_α as much as 1.960

Fourth: Define the decision-making criteria as it has been discussed above

Fifth: Make a decision of hypothesis testing results
The calculation result with IBM SPSS shows the t_o as much as 0.421 is lower than t_α as much as 1.960; thus the decision is: accept H0 and reject H1. This means that there is no a linear relationship between a service quality exogenous variable and a customer satisfaction level endogenous variable. Because there is no a linear relationship between the two variables; then the service quality has no effect on the customer satisfaction level significantly. Thus, the coefficient value of Beta (in column **Standardized Coefficient Beta**) as much as 0.038 is insignificant. This proves from the significance value in the **Sig column** as much as 0.675 which is bigger than 0.05.

Relationship between a Complaint Handling Exogenous Variable and a Customer Satisfaction Level Endogenous Variable

To see whether there is a linear relationship between a complaint handling exogenous variable and a customer satisfaction endogenous variable we can perform the following analysis steps:

First: Determine the hypothesis to be tested
H0: There is no linear relationship between a complaint handling exogenous variable and a customer satisfaction endogenous variable
H1: There is linear relationship between a complaint handling exogenous variable and a customer satisfaction endogenous variable

Second: Calculate the value of observation t (t_o)
The t_o value shown in column t in the above **Coefficients** table is 3.273
Third: Calculate the number of t table (t_α) as it has been discussed above
From these provisions it is obtained t_α as much as 1.960

Fourth: Define the decision-making criteria as it has been discussed above

Fifth: Make a decision of hypothesis testing results
The calculation result with IBM SPSS shows the t_o as much as 3.273 is bigger than t_α as much as 1.960; thus the decision is: reject H0 and accept H1. This means that there is a linear relationship between a complaint handling exogenous variable and a customer satisfaction level endogenous variable. Because there is a linear relationship between the two variables; then the complaint handling affects the customer satisfaction level significantly. The amount of effect can be known from coefficient value of Beta (in column **Standardized Coefficient Beta**) as much as 0.345. The effect of this magnitude is significant because the value of significance in the **Sig column** is 0.001 which is smaller than 0.05.

Correlation Among the Product Quality, Service Quality, and Complaint Handling Exogenous Variables

The magnitude of correlation among the exogenous variables of product quality, service quality, and complaint handling can be seen in the correlation table output.

Correlation Between Product Quality and Service Quality

To interpret correlation coefficient values are used the following criteria:

- 0: There is no correlation
- $> 0 - 0.25$: very weak correlation
- $> 0.25 - 0.5$: medium correlation
- $> 0.5 - 0.75$: strong correlation
- $> 0.75 - 0.99$: very strong correlation
- 1: perfect correlation

The correlation coefficient between product quality and service quality is 0.600. The correlation coefficient of 0.600 means that the relationship between those two variables is strong and positive. Positive correlation means if the variable of product quality is high then the service quality variable is also considered high and it applies also vice versa. The correlation of these two variables is significant because the significance level (Sig) of the study is 0.000 which is smaller than 0.05. The applicable provision is if the significance (sig) < 0.05 the correlation between the two variables is significant; otherwise if the significance (sig) > 0.05 the correlation between the two variables is not significant. Notes: when the output of SPSS for the correlation coefficients is given a two star mark (**), the significance level becomes 0.01 instead of 0.05.

Correlation Between Product Quality and Complaint Handling

The correlation coefficient between product quality and complaint handling is 0.724. The correlation coefficient of 0.724 means that the relationship between those two variables is strong and positive. The correlation of these two variables is significant because the significance level (Sig) is 0.000 which is smaller than 0.05.

Correlation Between Service Quality and Complaint Handling

The correlation coefficient between product quality and complaint handling is 0.684. The correlation coefficient of 0.684 means that the relationship between those two variables is strong and positive. The correlation of these two variables is significant because the significance level (Sig) is 0.000 which is smaller than 0.05.

5.6 Create a Path Diagram

After completion of the calculation and its interpretation then we must create a path diagram of the path analysis model we have made before completed with the values of each parameter. The path diagram will be as below:

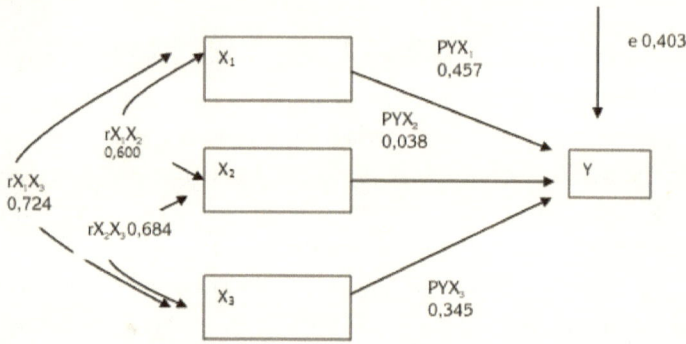

Figure 5.4 Path Diagram Parameter Values on the Effect of Product Quality, Service Quality and Handling Complaint Variables on Customer Satisfaction Level

Thus the structural equation for the path diagram above is:

$y = 0,457 X_1 + 0,038 X_2 + 0,345 X_3 + e$

5.7 Conclusion

The conclusion of the case above is as follows:

a. The effect of product quality on customer satisfaction is 0,457
b. The effect of service quality on customer satisfaction is 0,038
c. The effect of complaint handling on customer satisfaction is 0,345
d. Effect of product quality, service quality, and complaint handling on customer satisfaction of 0,597
e. The influence of other variables beyond this model is 0.403
f. The correlation between product quality and service quality is 0.600
g. The correlation between product quality and complaint handling is 0.724

h. The correlation between service quality and complaint handling is 0.684

5.8 Validity of Research Results

In this section, the discussion will focus on validity of the research result. Several matters to discuss are: Regression model eligibility, predictor accuracy, multicollinearity, linieariry, and normal distribution. We will discuss them one by one in the next session.

The Regression Model Eligibility

To find out if the above regression model is correct we will do the model testing. Model testing can be done in two ways, first using the value of F in ANOVA output table; the second by using the significance or probability value in the same output table

To perform hypothesis testing the steps are as follows

First: state the hypothesis
H0: There is no linear relationship among product quality, service quality, and complaint handling with customer satisfaction level.
H1: There is a linear relationship among product quality, service quality, and complaint handling with customer satisfaction level.

Second: Calculate the value of F research (F_o)
The value of F research from IBM SPSS output is 47.365

Third: Calculate F table (F α) with the following conditions: Determine the significance level of 0.05 and determine degree of freedom (DF) by calculating numerator / vector 1: number of variables -1 or 4 -1 = 3; and denominator / vector 2: number of cases - the number of variables or 100 - 4 = 96. With that provision it is obtained F table 2.68.

Fourth: Determining the decision-making criteria hypothesis test results as follows:
• If $F_o > F_\alpha$, then H0 is rejected and H1 accepted;
• If $F_o < F_\alpha$, then H0 is accepted and H1 is rejected

Fifth: Taking a decision on hypothesis test result
The calculation results F_o equal to 47,365 which is bigger than F_α equal to 2,68. Thus H0 is rejected and H1 is accepted. It means that there is a linear relationship between product quality, service quality, and complaint handling with customer satisfaction level. The conclusion is that the above regression model is eligible and correct.

Hypothesis Test Using Significance (sig) / P value

First: state the hypothesis

Second: Calculate the significance (sig) or probability value
The significance is 0.000

Third: Determine significance criteria (α) as much as 0.05 (Note: The default Sig of SPSS is 0.05)

Fourth: Determining the decision-making criteria
Criteria for decision-making is as follows:

- If significance < 0.05, then H0 is rejected and H1 is accepted
- If significance > 0.05, then H0 is accepted and H1 is rejected

Fifth: Make a decision

The calculation result the significance number as much as 0.000 which smaller than 0.05; thus H0 is rejected and H1 accepted. This means that there is a linear relationship among product quality, service quality, and complaint handling with customer satisfaction level. Accordingly, from the results of testing the above hypothesis proves that the regression model that we make is eligible.

Testing the Predictor Accuracy

To test the predictor accuracy (exogenous variables) used to predict the endogenous variables we can use standard deviation of each variable taken from Descriptive Output and standard error of estimate value taken from Model Summary Output. The provisions are as follows:

- If the value of the standard error of estimate <standard deviation; then

the predictor is feasible
- If the value of the standard error of estimate> standard deviation; then the predictor is not feasible

The calculation result above shows that the standard error of estimate value is 1,685 <standard deviation 3,135 (product quality variable), 2,903 (service quality variable) and 2,954 (complaint handling variable). In conclusion the three exogenous variables used as predictors are correct

Testing Multicollinearity

Multicolinilierity occurs in the independent variables (exogenous ones) if the correlation between independent variables is very high as much as 9 , or close to 1.

From the calculation result above correlation among exogenous variables is less than 0.9:

- Correlation between product quality and service quality is 0.600
- Correlation between product quality and complaint handling is 0.724
- Correlation between service quality and complaint handling is 0.684

Thus there is no multicollinearity among the exogenous variables used in this regression model.

Testing Linearity

To test linearity we use the Normal PP Plot of Regression Standardized Residual above. From the plot above it can be concluded the data has formed a straight line from the lower left to the top right in accordance with the theory of linearity, so it can be concluded that the linearity in this regression model has been met.

Testing Data Normality

To test data normality we use the histogram output above. The data is normally distributed if the data form a bell curve. If we see the picture above, although not its is perfect; data distribution shows tendency to form a bell curve; thus the data is considered to be normally distributed. The conclusion is that the regression model we have created already meets all the requirements. Note: another way to check the data distribution, we can use Kolmogorov – Smirnov procedure for the unstructured data and Jarque Berra for the panel data.

CHAPTER 6

AN APPLICATION OF A MEDIATION MODEL

6.1 A Case Example

In this mediation model we will use three independent variables that serve as exogenous variables and two dependent variables that serve as endogenous variables. As exogenous variables are product quality, service quality and complaint handling, while the first endogenous variable that serves as an intervening variable is customer satisfaction and the second endogenous variable is customer loyalty. The relationship between these variables when depicted in the path diagram model becomes as below:

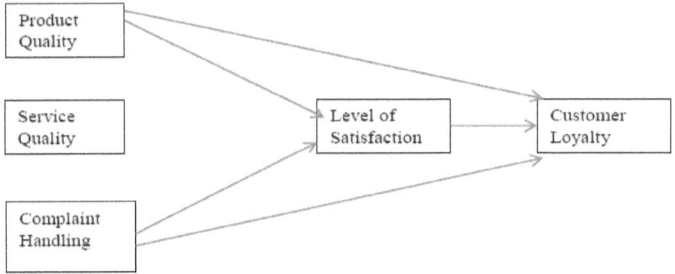

Figure 6.1 A Path Diagram Model on the Direct and Indirect Effect of Product Quality, Service Quality and Complaint Handling Variables on Customer Loyalty Through Customer Satisfaction Level Variable Based on Theory

6.2 Hypotheses and Problems

The first hypothesis for this case is:

H0: Variables of product quality, service quality and complaint handling have no effect on customer satisfaction level simultaneously and partially

H1: Variables of product quality, service quality and complaint handling have an effect on customer satisfaction level simultaneously and partially

The second hypothesis for this case is:

H0: Variables of product quality and complaint handling have no effect on customer loyalty simultaneously and partially

H1: Variables of product quality and complaint handling have an effect on customer loyalty simultaneously and partially

The third hypothesis for this case is:

H0: A customer satisfaction variable has no effect on a customer loyalty variable

H1: A customer satisfaction variable has an effect on a customer loyalty variable

The problem for this case is:

a. How much is the effect of product quality, service quality and complaint handling on customer satisfaction partially?

b. How much is the effect of product quality, service quality and complaint handling on customer satisfaction simultaneously?

c. How much is the effect of product quality, service quality, complaint handling and customer satisfaction on customer loyalty partially?

d. How much is the effect of product quality, service quality, complaint handling and customer satisfaction on customer loyalty simultaneously?

e. How much is the effect of product quality, service quality, complaint handling on customer loyalty through customer satisfaction?

6.3 Research Data

PATH ANALYSIS: DATA ANALYSIS APPLICATION

There are 100 data for the above case as follows:

No	product	service	complaint	satisfaction	loyalty	No	product	service	complaint	satisfaction	loyalty
1	18	17	16	15	14	51	13	14	13	12	13
2	15	17	18	12	11	52	13	13	14	10	9
3	17	14	16	14	11	53	15	16	16	15	14
4	14	14	14	13	10	54	12	12	14	12	11
5	15	15	16	12	11	55	13	12	15	9	9
6	17	15	16	13	11	56	15	15	16	14	12
7	13	16	12	14	13	57	14	18	12	10	9
8	19	19	20	11	12	58	13	11	13	11	10
9	15	16	17	14	14	59	10	11	13	12	11
10	19	19	18	14	15	60	16	18	16	12	11
11	14	16	17	12	11	61	14	11	11	10	9
12	15	11	15	10	9	62	17	12	12	13	12
13	14	14	13	12	11	63	16	14	13	12	11
14	16	16	18	11	10	64	14	15	12	10	9
15	10	14	16	12	11	65	13	11	14	10	9
16	13	15	17	13	12	66	13	11	14	11	10
17	19	12	17	14	12	67	12	11	14	9	9
18	15	12	16	14	12	68	13	14	13	10	9
19	15	12	14	15	12	69	15	12	13	9	9
20	16	14	15	14	12	70	13	13	17	12	12
21	12	11	15	14	11	71	13	14	12	11	10
22	18	16	14	14	13	72	13	16	15	12	12
23	13	15	15	11	11	73	15	14	15	12	12
24	14	14	13	12	11	74	14	13	15	10	9
25	13	13	17	11	10	75	14	14	13	10	10
26	10	10	12	9	9	76	16	16	18	15	16
27	13	17	13	9	9	77	20	19	22	17	16
28	12	12	11	10	9	78	17	20	20	16	15
29	11	14	12	8	6	79	16	20	20	15	15
30	9	14	8	7	8	80	18	16	16	16	16
31	10	10	13	12	11	81	15	18	19	12	12
32	8	12	12	9	8	82	16	19	19	12	12
33	13	14	12	10	9	83	20	16	15	13	13
34	8	15	14	9	8	84	21	22	22	19	18
35	11	8	12	10	9	85	19	19	19	12	12

47

36	13	13	16	11	12	86	18	15	16	13	13
37	11	14	11	9	9	87	18	19	20	12	12
38	9	15	14	11	10	88	17	15	18	12	12
39	15	11	12	12	11	89	21	17	20	17	18
40	11	12	15	8	9	90	16	17	19	17	16
41	11	12	13	8	8	91	16	19	18	16	16
42	9	14	14	9	9	92	19	16	20	16	15
43	10	17	11	9	9	93	17	19	18	16	14
44	9	10	12	8	8	94	19	18	21	13	12
45	12	9	12	7	9	95	21	17	20	19	18
46	13	15	15	8	8	96	19	16	17	15	15
47	13	8	14	10	9	97	19	15	19	15	15
48	14	11	11	9	9	98	17	19	19	15	14
49	11	13	10	9	8	99	19	19	19	12	13
50	12	13	15	13	12	100	18	17	19	14	15

6.4 Stages in Completing the Case

To solve the case above the stages are as follows:

First: Create a model of the path diagram based on the relationship among variables we are studying

The path diagram model of the relationship among variables we are studying is as follows:

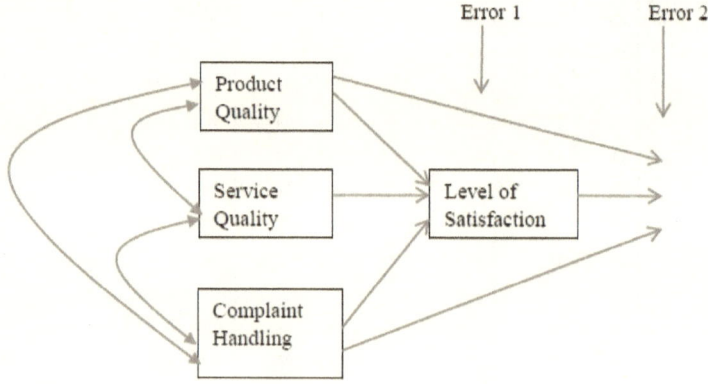

Figure 6.2 A Path Diagram Model on the Direct and Indirect Effect of Product Quality, Service Quality and Complaint Handling on Customer Loyalty Through Customer Satisfaction Level

PATH ANALYSIS: DATA ANALYSIS APPLICATION

Second: Create a path diagram of the above model

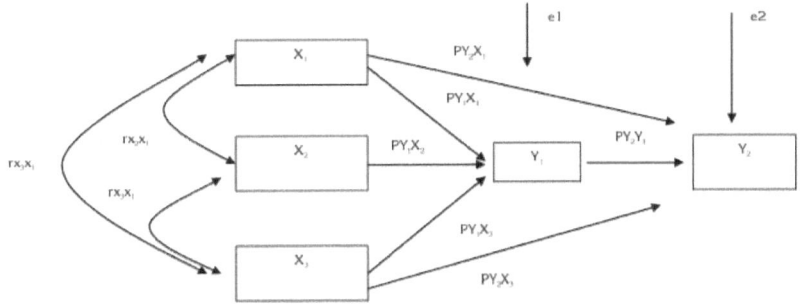

Figure 6.3 A Path Diagram on the Direct and Indirect Effect of Product Quality, Service Quality and Complaint Handling on Customer Loyalty Through Customer Satisfaction Level

Where:
X1 as a product quality exogenous variable
X2 as a service quality exogenous variable
X3 as a complaint handling exogenous variable
Y1 as a satisfaction level intervening endogenous variable
Y2 as a customer loyalty endogenous variable

Third: Create a structural equation
The path diagram above has two structural equations. The structural equation can be seen as below:

- $Y_1 = PY_1X_1 + PY_1X_2 + PY_1X_3 + e_1$ (structural equation of sub structure 1)

- $Y_2 = PY_2X_1 + PY_2Y_1 + PY_2X_3 + e_2$ (structural equation of sub structure 2)

Fourth: Creating a variable design, entering data and presenting it in IBM SPSS Design variables can be created by selecting the left bottom sub menu on the command: **Variable View**. Write down all variables pursuant to the example below

Name	Type	Width	Decimal	Label	Values	Missing	Column	Align	Measure	Role
product	Numeric	8	0	Product Quality	None	None	8	Right	Scale	Input
service	Numeric	8	0	Service Quality	None	None	8	Right	Scale	Input
complaint	Numeric	8	0	Complaint Handling	None	None	8	Right	Scale	Input
satisfaction	Numeric	8	0	Satisfaction level	None	None	8	Right	Scale	Both
loyalty	Numeric	8	0	Customer Loyalty	None	None	8	Right	Scale	Target

Fifth: Enter data by clicking on the command: **Data View**

No	product	service	complaint	satisfaction	loyalty
1					
.					
.					
100					

Sixth: Perform analysis by using IBM SPSS with the following steps
The calculation starts with Sub-Structure 1.

- Click **Analyze> Linear**
- Enter the satisfaction variable to the **Dependent** column
- Enter the product quality, service quality and complaint handling variables to **Independent** column
- **Statistics:** at the option of **Regression Coefficients**, check **Estimate, Model Fit** and **Descriptives > Continue**
- Click **Ok** to process

The result of the calculation for sub-structure 1 is as follows:

PATH ANALYSIS: DATA ANALYSIS APPLICATION

Output of the Regression Section

Model Summary[b]

Model	R	R Square	Adjusted R Square	Std. Error of the Estimate	Durbin-Watson
1	,773[a]	,597	,584	1,685	1,911

a. Predictors: (Constant), Product Quality, Service Quality, Complaint Handling
b. Dependent Variable: Customer Satisfaction

ANOVA[a]

Model		Sum of Squares	df	Mean Square	F	Sig.
1	Regression	403,413	3	134,471	47,365	,000[b]
	Residual	272,547	96	2,839		
	Total	675,960	99			

a. Dependent Variable: Customer Satisfaction
b. Predictors: (Constant), Complaint Handling, Service Quality, Product Quality

Coefficients[a]

Model		Unstandardized Coefficients		Standardized Coefficients	t	Sig.
		B	Std. Error	Beta		
1	(Constant)	1,327	,968		1,371	,173
	Product Quality	,381	,080	,457	4,758	,000
	Service Quality	,034	,082	,038	,421	,675
	Complaint Handling	,305	,093	,345	3,273	,001

a. Dependent Variable: Customer Satisfaction

Correlation Session

Correlations

		Customer Satisfaction	Product Quality	Service Quality	Complaint Handling
Pearson Correlation	Customer Satisfaction	1,000	,730	,548	,702
	Product Quality	,730	1,000	,600	,724
	Service Quality	,548	,600	1,000	,684
	Complaint Handling	,702	,724	,684	1,000
Sig. (1-tailed)	Customer Satisfaction	.	,000	,000	,000
	Product Quality	,000	.	,000	,000
	Service Quality	,000	,000	.	,000
	Complaint Handling	,000	,000	,000	.
N	Customer Satisfaction	100	100	100	100
	Product Quality	100	100	100	100
	Service Quality	100	100	100	100
	Complaint Handling	100	100	100	100

6.5 The interpretation of the calculation results for the Sub-Structure 1

In this section, the interpretation is divided into two: first, looking at the combined effect of the three exogenous variables on one endogenous variable, and second, looking at the partial effect for each of the exogenous variable on one endogenous variable

The Effect of Combined Product Quality, Service Quality, and Complaint Handling Exogenous Variables on Customer Satisfaction Level

The effect of product quality, service quality, and complaint handling exogenous variables on the customer satisfaction level variable can be seen in the output in the **Model Summary** table, in the R square value. The value of R square (R^2) in the table above is 0.597. This number shows the amount of variation of the customer satisfaction endogenous variable that can be explained by exogenous variables of product quality, service quality, and complaints handling. In another way, we can say the R square (R^2) of

0.597 shows the amount of the effect of exogenous variables of product quality, service quality, and complaints handling on the customer satisfaction endogenous variable.

This value can be made in the form of percent numbers by calculating Coefficient of Determination (CD) by using the following formula.

$CD = R^2 \times 100\%$

$CD = 0.597 \times 100\%$

$CD = 59.7\%$

While the rest can be calculated using the following formula:

$e = 1 - R^2$

$e = 1 - 0.597$

$e = 0.403$

The number 0.403 (40.3%) is the influence of other factors in the model beyond the three exogenous variables studied. In other words, the variation of customer satisfaction level which can be explained by using exogenous variables of product quality, service quality, and complaint handling is equal to 59.7%; while the 40.3% is caused by other variables outside of this research.

Simultaneous Hypothesis Testing to See The Effect of Exogenous Product Quality, Service Quality, and Complaint Handling Variables on a Customer Satisfaction Level Variable

This hypothesis testing is used to see the simultaneous effect of the product quality, service quality, and complaint handling variables on customer satisfaction level. The testing uses the value of significance (sig) from the ANOVA output table. To see whether there is a linear relationship between product quality, service quality, and complaint handling variables with customer satisfaction level we can perform the following analysis steps:

First: Determine the hypothesis to be tested

H0: There is no linear relationship between product quality, service quality, and complaint handling with customer satisfaction level

H1: There is linear relationship between product quality, service quality, and complaint handling with customer satisfaction level

Second: Calculate the value of significance (sig) or pvalue
The sig value shown in column Sig is 0.000

Third: use the following criteria
If the sig value < 0.05 reject H0 and accept H1
If the sig value > 0.05 accept H0 and reject H1

Fourth: take the decision of the hypothesis testing

Since the sig value is 0.000 < 0.05; therefore reject H0 and accept H1. Thus there is linear relationship between product quality, service quality, and combined handling with customer satisfaction level. Accordingly product quality, service quality, and complaint handling affect customer satisfaction level significantly.

The Partial Effect of Product Quality, Service Quality, and Complaint Handling Exogenous Variables on A Customer Satisfaction Level Endogenous Variable

The magnitude of the effect of exogenous independent variables of product quality, service quality, and complaint handling individually / partially on a customer satisfaction level endogenous variable can be seen from the Beta or Standardized Coefficients value. Whereas the testing of hypothesis uses t value. The values can be seen in the **Coefficient** table output. Therefore, in the following sections we will discuss the effect of exogenous variables of product quality, service quality, and complaint handling on a customer satisfaction endogenous variable partially.

Relationship between a Product Quality Exogenous Variable and a Customer Satisfaction Level Endogenous Variable

To see whether there is a linear relationship between a product quality exogenous variable and a customer satisfaction endogenous variable we can

PATH ANALYSIS: DATA ANALYSIS APPLICATION

perform the following analysis steps:

First: Determine the hypothesis to be tested
H0: There is no linear relationship between a product quality exogenous variable and a customer satisfaction endogenous variable
H1: There is linear relationship between a product quality exogenous variable and a customer satisfaction endogenous variable

Second: Calculate the value of observation t (t_o)
The t_o value shown in column t in the above **Coefficients** table is 4.758

Third: Calculate the number of t table (t_α) with the following provisions: Determine the level of significance of 0.05 and Degree of Freedom (DF) with the provisions: DF = n -2, or 100 - 2 = 98. From these provisions it is obtained t_α as much as 1.960

Fourth: Define the decision-making criteria as follows

If $t_o > t_\alpha$, then H0 is rejected and H1 accepted;
If $t_o < t_\alpha$, then H0 is accepted and H1 is rejected
or we can use the second alternative by using the significance level (probability / p value)
If sig < 0.05, then there is significant influence
If sig > 0.05, then there is no significant influence

Fifth: Make a decision of hypothesis testing results
The calculation result with IBM SPSS shows the t_o as much as 4.758 is bigger than t_α as much as 1.960; thus the decision is reject H0 and accept H1. This means that there is a linear relationship between a product quality exogenous variable and a customer satisfaction level endogenous variable. Because there is a linear relationship between the two variables; then the product quality affects the customer satisfaction level significantly. The amount of effect can be known from coefficient value of Beta (in column **Standardized Coefficient Beta**) as much as 0.457. The effect of this magnitude is significant because the value of significance in the **Sig column** is 0.000 which is smaller than 0.05.

Relationship between a Service Quality Exogenous Variable and a Customer Satisfaction Level Endogenous Variable

To see whether there is a linear relationship between a service quality exogenous variable and a customer satisfaction endogenous variable we can perform the following analysis steps:

First: Determine the hypothesis to be tested
H0: There is no linear relationship between a service quality exogenous variable and a customer satisfaction endogenous variable
H1: There is linear relationship between a service quality exogenous variable and a customer satisfaction endogenous variable

Second: Calculate the value of observation t (t_o)
The t_o value shown in column t in the above **Coefficients** table is 0,421

Third: Calculate the number of t table (t_α) as it has been discussed above
From these provisions it is obtained t_α as much as 1.960

Fourth: Define the decision-making criteria as it has been discussed above

Fifth: Make a decision of hypothesis testing results
The calculation result with IBM SPSS shows the t_o as much as 0.421 is lower than t_α as much as 1.960; thus the decision is accept H0 and reject H1. This means that there is no a linear relationship between a service quality exogenous variable and a customer satisfaction level endogenous variable. Because there is no a linear relationship between the two variables; then the service quality has no effect on the customer satisfaction level significantly. Thus, the coefficient value of Beta (in column **Standardized Coefficient Beta**) as much as 0.038 is insignificant. This proves from the significance value in the **Sig column** as much as 0.675 which is bigger than 0.05.

Relationship between a Complaint Handling Exogenous Variable and a Customer Satisfaction Level Endogenous Variable

To see whether there is a linear relationship between a complaint handling exogenous variable and a customer satisfaction endogenous variable we can

PATH ANALYSIS: DATA ANALYSIS APPLICATION

perform the following analysis steps:

First: Determine the hypothesis to be tested
H0: There is no linear relationship between a complaint handling exogenous variable and a customer satisfaction endogenous variable
H1: There is linear relationship between a complaint handling exogenous variable and a customer satisfaction endogenous variable

Second: Calculate the value of observation t (t_o)
The t_o value shown in column t in the above **Coefficients** table is 3.273

Third: Calculate the number of t table (t_α) as it has been discussed above
From these provisions it is obtained t_α as much as 1.960

Fourth: Define the decision-making criteria it as has been discussed above

Fifth: Make a decision of hypothesis testing results
The calculation result with IBM SPSS shows the t_o as much as 3.273 is bigger than t_α as much as 1.960; thus the decision is reject H0 and accept H1. This means that there is a linear relationship between a complaint handling exogenous variable and a customer satisfaction level endogenous variable. Because there is a linear relationship between the two variables; then the complaint handling affects the customer satisfaction level significantly. The amount of effect can be known from coefficient value of Beta (in column **Standardized Coefficient Beta**) as much as 0.345. The effect of this magnitude is significant because the value of significance in the **Sig column** is 0.001 which is smaller than 0.05.

Correlation Among the Product Quality, Service Quality, and Complaint Handling Exogenous Variables

The magnitude of correlation among the exogenous variables of product quality, service quality, and complaint handling can be seen in the correlation table output.

Correlation Between Product Quality and Service Quality

The correlation coefficient between product quality and service quality is 0.600. The correlation coefficient of 0.600 means that the relationship between those two variables is strong and positive. Positive correlation means if the variable of product quality is high then the service quality variable is also considered high and it applies also vice versa. The correlation of these two variables is significant because the significance level (Sig) of the study is 0.000 which is smaller than 0.05. The applicable provision is if the significance (sig) < 0.05 the correlation between the two variables is significant; otherwise if the significance (sig) > 0.05 the correlation between the two variables is not significant. Notes: when the output of SPSS for the correlation coefficients is given a two star mark (**), the significance level becomes 0.01 instead of 0.05.

Correlation Between Product Quality and Complaint Handling
The correlation coefficient between product quality and complaint handling is 0.724. The correlation coefficient of 0.724 means that the relationship between those two variables is strong and positive. The correlation of these two variables is significant because the significance level (Sig) is 0.000 which is smaller than 0.05.

Correlation Between Service Quality and Complaint Handling
The correlation coefficient between product quality and complaint handling is 0.684. The correlation coefficient of 0.684 means that the relationship between those two variables is strong and positive. The correlation of these two variables is significant because the significance level (Sig) is 0.000 which is smaller than 0.05.

Notes: To conduct the validity results in this model, refer to the previous chapter's discussion (Chapter 5: 5.8 session)

6.6 Calculation and Interpretation of the Sub-Structure 2
Use the following steps to calculate for the sub structure 2

- Click **Analyze> Linear**
- Enter the loyalty variable to the **Dependent** column
- Enter the product quality, service quality, complaint handling and satisfaction variables to **Independent** column
- **Statistics:** at the option of **Regression Coefficients**, check **Estimate, Model Fit** and **Descriptives > Continue** Click **Ok** to process

The result is as follows:

Model Summary

Model	R	R Square	Adjusted R Square	Std. Error of the Estimate
1	,947a	,896	,892	,838

a. Predictors: (Constant), Customer Satisfaction, Service Quality, Product Quality, Complaint Handling

ANOVAa

Model		Sum of Squares	Df	Mean Square	F	Sig.
1	Regression	577,315	4	144,329	205,613	,000b
	Residual	66,685	95	,702		
	Total	644,000	99			

a. Dependent Variable: Customer Loyalty
b. Predictors: (Constant), Customer Satisfaction, Service Quality, Product Quality, Complaint Handling

Coefficientsa

Model		Unstandardized Coefficients		Standardized Coefficients	t	Sig.
		B	Std. Error	Beta		
1	(Constant)	-,847	,486		-1,744	,084
	Product Quality	,068	,044	,083	1,531	,129
	Service Quality	,073	,041	,083	1,787	,077
	Complaint Handling	,086	,049	,099	1,757	,082
	Customer Satisfaction	,740	,051	,758	14,578	,000

a. Dependent Variable: Customer Loyalty

Interpretation of the Results

In this section, the interpretation is divided into two: first, looking at the combined effect of the four exogenous variables on one endogenous variable, and second, looking at the partial effect for each of the exogenous variable on one endogenous variable.

The Effect of Combined Exogenous Product Quality, Service Quality, and Complaint Handling and Customer Satisfaction Level Variables on a Loyalty Variable

The effect of product quality, service quality, complaint handling and customer satisfaction level exogenous variables on the loyalty endogenous variable can be seen in the output in the **Model Summary** table, in the R square value. The value of R square (R^2) in the table above is 0.896. This number shows the amount of variation of the loyalty endogenous variable that can be explained by exogenous variables of product quality, service quality, complaints handling and customer satisfaction. In another way, we can say the R square (R^2) of 0.896 shows the amount of the effect of exogenous variables of product quality, service quality, complaint handling and customer satisfaction on the loyalty endogenous variable.

This value can be made in the form of percent numbers by calculating Coefficient of Determination (CD) by using the following formula.

CD = R^2 x 100%

CD = 0.896 x 100%

CD = 89.6%

While the rest can be calculated using the following formula:

e = 1 - R^2

e = 1 – 0.896

e = 0.104

The number 0.104 (10.4%) is the influence of other factors in the model beyond the four exogenous variables studied. In other words, the variation of loyalty which can be explained by using exogenous variables of product quality, service quality, complaint handling and customer satisfaction level is equal to 89.6%; while the 10.4% is caused by other variables outside of this research.

Simultaneous Hypothesis Testing to See The Effect of Exogenous Product Quality, Service Quality, Complaint Handling and Customer Satisfaction Level Variables on a Customer Loyalty Variable

This hypothesis testing is used to see the simultaneous effect of the product quality, service quality, complaint handling and customer satisfaction level on customer loyalty. The testing uses the value of significance (sig) from the ANOVA output table. To see whether there is a linear relationship between product quality, service quality, and combined complaint handling variables with customer satisfaction level we can perform the following analysis steps:

First: Determine the hypothesis to be tested
H0: There is no linear relationship between the product quality, service quality, complaint handling and customer satisfaction level with customer loyalty
H1: There is linear relationship between the product quality, service quality, complaint handling and customer satisfaction level with customer loyalty

Second: Calculate the value of significance (sig) or pvalue
The sig value shown in column Sig is 0.000
Third: use the following criteria
If the sig value < 0.05 reject H0 and accept H1
If the sig value > 0.05 accept H0 and reject H1

Fourth: take the decision of the hypothesis testing

Since the sig value is 0.000 < 0.05; therefore reject H0 and accept H1. Thus there is linear relationship between product quality, service quality, complaint handling and customer satisfaction level with customer loyalty. Accordingly product quality, service quality, complaint handling and customer satisfaction level affect customer loyalty significantly.

The Partial Effect of Product Quality, Service Quality, Complaint Handling and Customer Satisfaction Level Exogenous Variables on A Loyalty Endogenous Variable

The magnitude of the effect of exogenous variables of product quality, service quality, complaint handling and customer satisfaction level individually / partially on a loyalty endogenous variable can be seen from the Beta or Standardized Coefficients value. Whereas the testing of hypothesis uses t value. The values can be seen in the **Coefficient** table output. Therefore, in the following sections we will discuss the effect of exogenous variables of product quality, service quality, complaint handling and customer satisfaction on a loyalty endogenous variable partially.

Relationship between a Product Quality Exogenous Variable and a Loyalty Endogenous Variable

To see whether there is a linear relationship between a product quality exogenous variable and a loyalty endogenous variable we can perform the following analysis steps:

First: Determine the hypothesis to be tested
H0: There is no linear relationship between a product quality exogenous variable and a loyalty endogenous variable
H1: There is linear relationship between a product quality exogenous variable and a loyalty endogenous variable

Second: Calculate the value of observation t (t_o)
The t_o value shown in column t in the above **Coefficients** table is 1.531

Third: Calculate the number of t table (t_α) with the following provisions: Determine the level of significance of 0.05 and Degree of Freedom (DF) with the provisions: DF = n -2, or 100 - 2 = 98. From these provisions it is obtained t_α as much as 1.960

Fourth: Define the decision-making criteria as follows
If $t_o > t_\alpha$, then H0 is rejected and H1 accepted;
If $t_o < t_\alpha$, then H0 is accepted and H1 is rejected
or we can use the second alternative by using the significance level

PATH ANALYSIS: DATA ANALYSIS APPLICATION

(probability / p value)
If sig < 0.05, then there is significant influence
If sig > 0.05, then there is no significant influence

Fifth: Make a decision of hypothesis testing results
The calculation result with IBM SPSS shows the t_o as much as 1.531 is less than t_α as much as 1.960; thus the decision is: accept H0 and reject H1. This means that there is no a linear relationship between a product quality exogenous variable and a loyalty endogenous variable. Because there is no a linear relationship between the two variables; then the product quality does not affect the loyalty variable significantly. Thus, the coefficient value of Beta (in column **Standardized Coefficient Beta**) as much as 0.083 is insignificant. This proves from the significance value in the **Sig column** as much as 0.129 which is bigger than 0.05.

Relationship between a Service Quality Exogenous Variable and a Loyalty Endogenous Variable

To see whether there is a linear relationship between a service quality exogenous variable and a loyalty endogenous variable we can perform the following analysis steps:

First: Determine the hypothesis to be tested
H0: There is no linear relationship between a service quality exogenous variable and a loyalty endogenous variable
H1: There is linear relationship between a service quality exogenous variable and a loyalty endogenous variable

Second: Calculate the value of observation t (t_o)
The t_o value shown in column t in the above **Coefficients** table is 1.787

Third: Calculate the number of t table (t_α) as it has been discussed above
The t table is as much as 1.96
Fourth: Define the decision-making criteria as it has been discussed above

Fifth: Make a decision of hypothesis testing results
The calculation result with IBM SPSS shows the t_o as much as 1.787 is less

than t_α as much as 1.960; thus the decision: is accept H0 and reject H1. This means that there is no a linear relationship between a service quality exogenous variable and a loyalty endogenous variable. Because there is no a linear relationship between the two variables; then the service quality does not affect the loyalty variable significantly. Thus, the coefficient value of Beta (in column **Standardized Coefficient Beta**) as much as 0.083 is insignificant. This proves from the significance value in the **Sig column** as much as 0.077 which is bigger than 0.05.

Relationship between a Complaint Handling Exogenous Variable and a Loyalty Endogenous Variable

To see whether there is a linear relationship between a complaint handling exogenous variable and a loyalty endogenous variable we can perform the following analysis steps:

First: Determine the hypothesis to be tested
H0: There is no linear relationship between a complaint handling exogenous variable and a loyalty endogenous variable
H1: There is linear relationship between a complaint handling exogenous variable and a loyalty endogenous variable

Second: Calculate the value of observation t (t_o)
The t_o value shown in column t in the above **Coefficients** table is 1.757

Third: Calculate the number of t table (t_α) as it has been discussed above
The t table is as much as 1.96

Fourth: Define the decision-making criteria as it has been discussed above

Fifth: Make a decision of hypothesis testing results
The calculation result with IBM SPSS shows the t_o as much as 1.757 is less than t_α as much as 1.960; thus the decision is: accept H0 and reject H1. This means that there is no a linear relationship between a complaint handling exogenous variable and a loyalty endogenous variable. Because there is no a linear relationship between the two variables; then the complaint handling does not affect the loyalty variable significantly. Thus, the coefficient value of Beta (in column **Standardized Coefficient Beta**) as much as 0.099 is insignificant. This proves from the significance value in the **Sig column** as much as 0.082 which is bigger than 0.05.

PATH ANALYSIS: DATA ANALYSIS APPLICATION

Relationship between a Customer Satisfaction Exogenous Variable and a Loyalty Endogenous Variable

To see whether there is a linear relationship between a customer satisfaction exogenous variable and a loyalty endogenous variable we can perform the following analysis steps:

First: Determine the hypothesis to be tested
H0: There is no linear relationship between a customer satisfaction exogenous variable and a loyalty endogenous variable
H1: There is linear relationship between a customer satisfaction exogenous variable and a loyalty endogenous variable

Second: Calculate the value of observation t (t_o)
The t_o value shown in column t in the above **Coefficients** table is 14.578

Third: Calculate the number of t table (t_α) as it has been discussed above
The t table is as much as 1.96

Fourth: Define the decision-making criteria as it has been discussed above

Fifth: Make a decision of hypothesis testing results
The calculation result with IBM SPSS shows the t_o as much as 14.578 is bigger than t_α as much as 1.960; thus the decision is: reject H0 and accept H1. This means that there is a linear relationship between a customer satisfaction exogenous variable and a loyalty endogenous variable. Because there is no a linear relationship between the two variables; then the customer satisfaction affects the loyalty variable significantly. The amount of effect can be known from coefficient value of Beta (in column **Standardized Coefficient Beta**) as much as 0,758. The effect of this magnitude is significant because the value of significance in the **Sig column** is 0.000 which is smaller than 0.05

Correlation Among the Product Quality, Service Quality, Complaint Handling and Customer Satisfaction Exogenous Variables

To calculate correlation is as follows:
- **Analyse > Correlate > Bivariate**
- Move variables of product quality, service quality, complaint handling and customer satisfaction from the left column to the **Variables** column; check **Pearson for Correlation Coefficients** option
- Click **OK**

The result is as follows

Correlations

		Product Quality	Service Quality	Complaint Handling	Customer Satisfaction
Product Quality	Pearson Correlation	1	,600"	,724"	,730"
	Sig. (2-tailed)		,000	,000	,000
	N	100	100	100	100
Service Quality	Pearson Correlation	,600"	1	,684"	,548"
	Sig. (2-tailed)	,000		,000	,000
	N	100	100	100	100
Complaint Handling	Pearson Correlation	,724"	,684"	1	,702"
	Sig. (2-tailed)	,000	,000		,000
	N	100	100	100	100
Customer Satisfaction	Pearson Correlation	,730"	,548"	,702"	1
	Sig. (2-tailed)	,000	,000	,000	
	N	100	100	100	100

**. Correlation is significant at the 0.01 level (2-tailed).

The magnitude of correlation among the exogenous variables of product quality, service quality, complaint handling and customer satisfaction can be seen in the correlation table output above.

Correlation Between Product Quality and Service Quality

The correlation coefficient between product quality and service quality is 0.600. The correlation coefficient of 0.600 means that the relationship between those two variables is strong and positive. Positive correlation means if the variable of product quality is high then the service quality variable is also considered high and it applies also vice versa. The correlation of these two variables is significant because the significance level (Sig) is 0.000 which is smaller than 0.01.

Correlation Between Product Quality and Complaint Handling

The correlation coefficient between product quality and complaint handling is 0.724. The correlation coefficient of 0.724 means that the relationship between those two variables is strong and positive. The correlation of these two variables is significant because the significance level (Sig) is 0.000 which is smaller than 0.01.

Correlation Between Service Quality and Complaint Handling
The correlation coefficient between product quality and complaint handling is 0.684. The correlation coefficient of 0.684 means that the relationship between those two variables is strong and positive. The correlation of these two variables is significant because the significance level (Sig) is 0.000 which is smaller than 0.01.

6.7 Create a Path Diagram

After completion of the calculation and its interpretation then we must create a path diagram of the path analysis model we have made before completed with the values of each parameter. The path diagram will be as below:

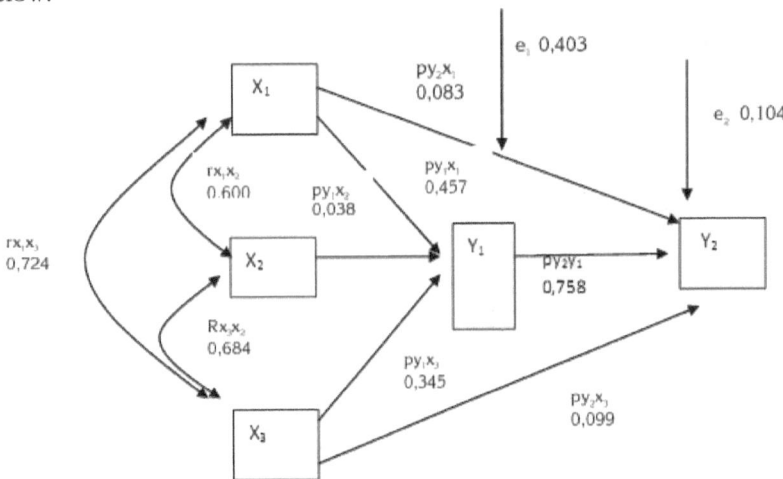

Figure 6.4 Path Diagram of Parameter Values on the Effect of Product Quality, Service Quality, Complaint Handling and Customer Satisfaction Level Variables on Customer Loyalty

The path diagram above has the following structural equations:
Substructure 1: $Y1 = 0.457 X + 0.038X + 0.345X + e_1$

Substructure 2: $Y2 = 0.083 X + 0.758 Y + 0.094 X + e_2$

Calculating Direct Effects (DE)
To calculate the direct effects (DE) is as follows:
- Effect of product quality on customer satisfaction
  X1 to Y1 = 0.457
- Effect of service quality to customer satisfaction
  X2 to Y1 = 0.038
- Effect of complaint handling to customer satisfaction
  X3 to Y1 = 0.345
- Effect of product quality on customer loyalty
  X1 to Y2 = 0.083
- Effect of service quality on loyalty
  X3 to Y2 = 0.083
- Effect of complaint handling on loyalty
  X3 to Y2 = 0.099
- Effect of customer satisfaction on loyalty
  Y1 to Y2 = 0.758

Calculating Indirect Effect (IE)
To calculate the indirect effects (IE) is as follows:
- Effect of product quality on customer loyalty through customer satisfaction
  PY1X1 x PY2Y1 = (0.457 x 0.758) = 0.346
- Effect of service quality on customer loyalty through customer satisfaction
  PY1X2 x PY2Y1 = (0,038 x 0,758) = 0,028
- Effect of complaint handling on customer loyalty through customer satisfaction
  PY1X3 x PY2Y1 = (0.345 x 0.758) = 0.261

Calculating the Total Effect (TE)
To calculate the total effect (TE) is as follows:
- Effect of product quality variable on customer loyalty through customer satisfaction
  PY1X1 + PY2Y1 = (0,457 + 0,758) = 1,215
- Effect of service quality on customer loyalty through customer satisfaction
  PY1X2 + PY2Y1 = (0,038 + 0,758) = 0,796

PATH ANALYSIS: DATA ANALYSIS APPLICATION

- Effect of complaint handling on customer loyalty through customer satisfaction
  PY1X3 + PY2Y1 = (0.345 + 0.758) = 1.103

6.8 Conclusion of Mediation Model
Based on the results of the calculation analysis above, we can draw the following conclusions:
a) Direct effect of product quality on customer loyalty is 0,083
b) Direct effect of service quality on customer loyalty is 0,083.
c) Direct effect of complaint handling on customer loyalty is 0,094.
d) Direct effect of customer satisfaction on customer loyalty is 0.758.
e) Combined effect of product quality, service quality, complaints handling, and customer satisfaction on customer loyalty is 0.896.
f) Effect of other variables beyond this model on customer loyalty is 0.104.
g) Direct effect of product quality on customer satisfaction is 0.457.
h) Direct effect of service quality on customer satisfaction is 0.038.
i) Direct effect of complaint handling to customer satisfaction is 0.345.
j) Combined effect of product quality, service quality, and complaint handling on customer satisfaction is 0,597.
k) Effect of other variables beyond the model on customer satisfaction is 0.403.
l) Effect of product quality on customer loyalty through customer satisfaction is 0.346
m) Effect of service quality on customer loyalty through customer satisfaction is 0,028
n) Effect of complaint handling on customer loyalty through customer satisfaction is 0.261

CHAPTER 7

AN APPLICATION OF A JOINT MULTIPLE REGRESSION AND MEDIATION MODEL

7.1 A Case Example

In this example of the joint (combined) multiple regression and mediation model, we will use two independent variables that serve as exogenous variables and two dependent variables that serve as endogenous variables. As exogenous variables are complaint handling and service quality, while the first endogenous variable that serves as an intervening variable is customer satisfaction and the second endogenous variable is customer loyalty.

The difference with the previous model is that there is no association relationship between the first exogenous variable (complaint handling) with the second exogenous variable (service quality); instead the causal relationship, namely the variable complaint handling gives effect to the variable of service quality. Thus the service quality variable serves as an endogenous variable in this context. This difference will become clear if we look at the path diagram model as follows:

PATH ANALYSIS: DATA ANALYSIS APPLICATION

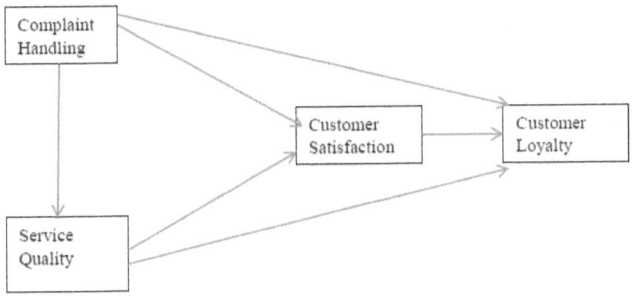

Figure 7.1 A Path Diagram Model of a Joint Multiple Regression and Mediation Model on Relationship among Complaint Handling, Service Quality, Customer Satisfaction and Customer Loyalty Variables Based on Theory

The model above can be broken down into 3 parts, as follows:
• First: The complaint handling variable directly affects the service quality
• Second: Variables of complaint handling and service quality affect both partially and simultaneously on customer satisfaction
• Third: Variables of complaint handling, service quality and customer satisfaction affect both partially and simultaneously on customer loyalty

7.2 Hypotheses and Problems

The first hypothesis for this case is:
H0: The complaint handling variable has no effect on service quality
H1: The complaint handling variable affects service quality
The second hypothesis for this case is:
H0: Variables of complaint handling and service quality have no effect on customer satisfaction either simultaneously or partially
H1: Variables of complaint handling and service quality have effect on customer satisfaction either simultaneously or partially
The third hypothesis for this case is:
H0: Variables of complaint handling and service quality through the customer satisfaction variable do not affect customer loyalty either simultaneously or partially
H1: Variables of complaint handling and service quality through the

customer satisfaction variable affect customer loyalty either simultaneously or partially

The problem for this case is:
a. How much does the variable of complaint handling affect service quality?
b. How much do the variables of complaint handling and service quality affect customer satisfaction simultaneously or partially?
c. How much do the variables of complaint handling, service quality and customer satisfaction affect on customer loyalty partially?
d. How much do the variables of complaint handling, service quality and customer satisfaction affect on customer loyalty simultaneously?
e. How much do variables of complaint handling and service quality through customer satisfaction affect on customer loyalty?

7.3 Research Data
There are 100 data for the above case as follows:

PATH ANALYSIS: DATA ANALYSIS APPLICATION

No	Complaint	Service	Satisfaction	Loyalty	No	Complaint	Service	Satisfaction	Loyalty
1	18	17	16	15	51	13	14	13	12
2	15	17	18	12	52	13	13	14	10
3	17	14	16	14	53	15	16	16	15
4	14	14	14	13	54	12	12	14	12
5	15	15	16	12	55	13	12	15	9
6	17	15	16	13	56	15	15	16	14
7	13	16	12	14	57	14	18	12	10
8	19	19	20	11	58	13	11	13	11
9	15	16	17	14	59	10	11	13	12
10	19	19	18	14	60	16	18	16	12
11	14	16	17	12	61	14	11	11	10
12	15	11	15	10	62	17	12	12	13
13	14	14	13	12	63	16	14	13	12
14	16	16	18	11	64	14	15	12	10
15	10	14	16	12	65	13	11	14	10
16	13	15	17	13	66	13	11	14	11
17	19	12	17	14	67	12	11	14	9
18	15	12	16	14	68	13	14	13	10
19	15	12	14	15	69	15	12	13	9
20	16	14	15	14	70	13	13	17	12
21	12	11	15	14	71	13	14	12	11
22	18	16	14	14	72	13	16	15	12
23	13	15	15	11	73	15	14	15	12
24	14	14	13	12	74	14	13	15	10
25	13	13	17	11	75	14	14	13	10
26	10	10	12	9	76	16	16	18	15

7.4 Stages in Completing the Case

To solve the case the stages are as below:

First: Creating a path diagram model based on the relationship between the variables we are examining:

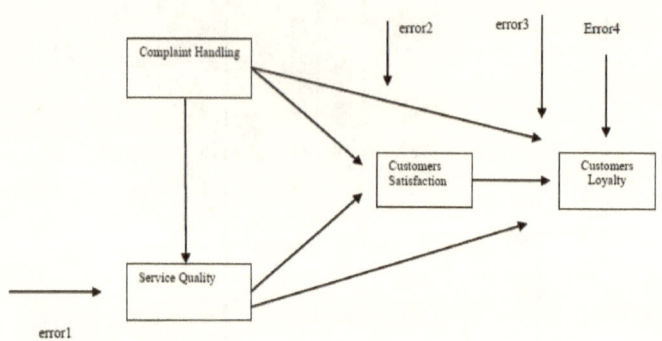

Figure 7.2 A Path Diagram Model of a Joint Multiple Regression and Mediation Model on Relationship among Complaint Handling, Service Quality, Customer Satisfaction and Customer Loyalty Variables

Second: Create a path diagram of the above model

PATH ANALYSIS: DATA ANALYSIS APPLICATION

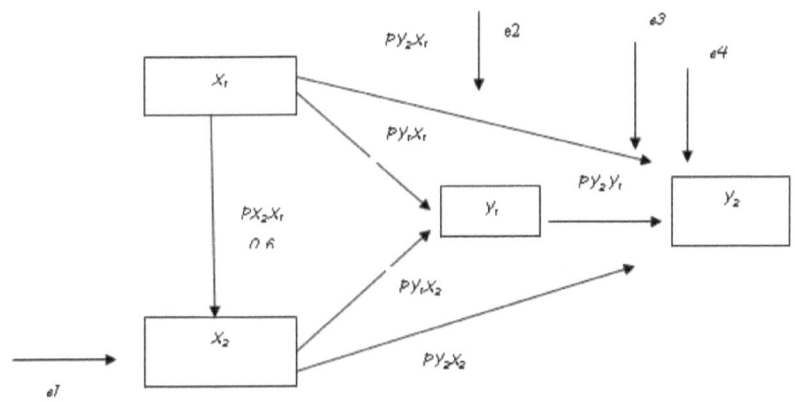

Figure 7.3 A Path Diagram of a Joint Multiple Regression and Mediation Model on Relationship among Complaint Handling, Service Quality, Customer Satisfaction and Customer Loyalty Variables

Where:
X1 as a complaint handling exogenous variable
X2 as a service quality endogenous variable
Y1 as a customer satisfaction endogenous intervening variable
Y2 as a customer loyalty endogenous variable

Third: Create a structural equation
The path diagram above has four structural equations. The structural equation can be seen as below:

- X2 = PX2X1 + e1 (As a structural equation of sub structure 1)
- Y1 = PY1X1 + PY1 X2 + e2 (As a structural equation of sub structure 2)
- Y2 = PY2 Y1 + e3 (As a structural equation of sub structure 3)
- Y2 = PY2X1 + PY2 X2 + e4 (As a structural equation of sub structure 4)

Fourth: Creating a variable design, entering data and presenting it in IBM SPSS

Design variables can be created by selecting the left bottom sub menu on the command: **Variable View**. Write down all variables pursuant to the example below

Name	Type	Width	Decimal	Label	Values	Missing	Column	Align	Measure	Role
complaint	Numeric	8	0	Complaint Handling	None	None	8	Right	Scale	Input
service	Numeric	8	0	Service Quality	None	None	8	Right	Scale	Input
satisfaction	Numeric	8	0	Satisfaction level	None	None	8	Right	Scale	Both
loyalty	Numeric	8	0	Customer Loyalty	None	None	8	Right	Scale	Target

Fifth: Enter data by clicking on the command: Data View

No	complaint	service	satisfaction	loyalty
1				
.				
.				
100				

Sixth: Perform analysis by using IBM SPSS with the following steps

Sub Structure 1

- Click **Analyze> Linear**
- Enter the service quality variable to the **Dependent** column
- Enter the complaint handling variable to **Independent** column
- **Statistics:** at the option of **Regression Coefficients**, check **Estimate, Model Fit** and **Descriptives > Continue**
- Click **Ok** to process

The result of calculation is as follows:

Model Summary

Model	R	R Square	Adjusted R Square	Std. Error of the Estimate
1	,600a	,360	,354	2,334

a. Predictors: (Constant), Complaint Handling

ANOVAa

Model		Sum of Squares	Df	Mean Square	F	Sig.
1	Regression	300,637	1	300,637	55,202	,000b
	Residual	533,723	98	5,446		
	Total	834,360	99			

a. Dependent Variable: Service Quality
b. Predictors: (Constant), Complaint Handling

Coefficientsa

Model		Unstandardized Coefficients		Standardized Coefficients	t	Sig.
		B	Std. Error	Beta		
1	(Constant)	6,509	1,111		5,858	,000
	Complaint Handling	,556	,075	,600	7,430	,000

a. Dependent Variable: Service Quality

The Interpretation of the Calculation Result for the Sub Structure 1

The Effect of an Exogenous Complaint Handling Variable on a Service Quality Variable

The effect of the complaint handling exogenous variable on the service quality variable can be seen in the output in the **Model Summary** table, in the R square value. The value of R square (R^2) in the table above is 0.360. This number shows the amount of variation of the service quality endogenous variable that can be explained by an exogenous variable of complaints handling. In another way, we can say the R square (R^2) of 0.360 shows the amount of the effect of the exogenous variable of complaint handling on the service quality endogenous variable.

This value can be made in the form of percent numbers by calculating Coefficient of Determination (CD) by using the following formula.

$CD = R^2 \times 100\%$

$CD = 0{,}360 \times 100\%$

$CD = 36\%$

While the rest can be calculated using the following formula:

$e = 1 - R^2$

$e = 1 - 0.36$

$e = 0.64$

The number 0.64 (64%) is the influence of other factors in the model beyond the complaint handling exogenous variable studied. In other words, the variation of service quality which can be explained by using the exogenous variable of complaint handling is equal to 36%; while the 64% is caused by other variables outside of this research

Hypothesis Testing to See The Effect of a Complaint Handling Variable on a Service Quality Variable

This hypothesis testing is used to see the effect of the complaint handling on service quality. The testing uses the value of significance (sig) from the ANOVA output table. To see whether there is a linear relationship between complaint handling and service quality we can perform the following analysis steps:

First: Determine the hypothesis to be tested
H0: There is no linear relationship between complaint handling and service quality
H1: There is linear relationship between complaint handling and service quality

Second: Calculate the value of significance (sig) or pvalue
The sig value shown in column Sig is 0.000

PATH ANALYSIS: DATA ANALYSIS APPLICATION

Third: use the following criteria
If the sig value < 0.05 reject H0 and accept H1
If the sig value > 0.05 accept H0 and reject H1

Fourth: take the decision of the hypothesis testing

Since the sig value is 0.000 < 0.05; therefore reject H0 and accept H1. Thus there is linear relationship between complaint handling and service quality. Accordingly complaint handling affects service quality significantly.

Relationship between a Complaint Handling Exogenous Variable and a Service Quality Endogenous Variable

To see whether there is a linear relationship between a complaint handling exogenous variable and a service quality endogenous variable we can perform the following analysis steps:

First: Determine the hypothesis to be tested
H0: There is no linear relationship between a complaint handling exogenous variable and a service quality variable
H1: There is linear relationship between a complaint handling exogenous variable and a service quality endogenous variable

Second: Calculate the value of observation t (t_o)
The t_o value shown in column t in the above **Coefficients** table is 7.430

Third: Calculate the number of t table (t_α) as it has been discussed in the previous chapter
From the provisions it is obtained t_α as much as 1.960

Fourth: Define the decision-making criteria as it has been discussed in the previous chapter

Fifth: Make a decision of hypothesis testing results
The calculation result with IBM SPSS shows the t_o as much as 7.430 is bigger than t_α as much as 1.960; thus the decision is reject H0 and accept H1. This means that there is a linear relationship between a complaint handling exogenous variable and a service quality endogenous variable.

Because there is a linear relationship between the two variables; then the complaint handling affects the service quality significantly. The amount of effect can be known from coefficient value of Beta (in column **Standardized Coefficient Beta**) as much as 0.600. The effect of this magnitude is significant because the value of significance in the **Sig column** is 0.000 which is smaller than 0.05.

Sub-Structure 2

To calculate the sub – structure 2, the steps are as follows:

- Click **Analyze> Linear**
- Enter the customer satisfaction variable to the **Dependent** column
- Enter the complaint handling and service quality variables to **Independent** column
- **Statistics:** at the option of **Regression Coefficients**, check **Estimate, Model Fit** and **Descriptives > Continue**
- Click **Ok** to process

The result of the calculation is as follows:

Model Summary

Model	R	R Square	Adjusted R Square	Std. Error of the Estimate
1	,788a	,621	,614	1,836

a. Predictors: (Constant), Service Quality, Complaint Handling

ANOVAa

Model		Sum of Squares	Df	Mean Square	F	Sig.
1	Regression	536,719	2	268,360	79,607	,000b
	Residual	326,991	97	3,371		
	Total	863,710	99			

a. Dependent Variable: Customer Satisfaction
b. Predictors: (Constant), Service Quality, Complaint Handling

PATH ANALYSIS: DATA ANALYSIS APPLICATION

Coefficients[a]

Model		Unstandardized Coefficients		Standardized Coefficients	t	Sig.
		B	Std. Error	Beta		
1	(Constant)	2,785	1,016		2,742	,007
	Complaint Handling	,462	,074	,491	6,283	,000
	Service Quality	,396	,079	,389	4,981	,000

a. Dependent Variable: Customer Satisfaction

The Interpretation of the Calculation Result for the Sub Structure 2

The Combined Effect of Service Quality and Complaint Handling Exogenous Variables on Customer Satisfaction Level

The effect of service quality and complaint handling exogenous variables on the customer satisfaction level variable can be seen in the output in the **Model Summary** table, in the R square value. The value of R square (R^2) in the table above is 0.621. This number shows the amount of variation of the customer satisfaction endogenous variable that can be explained by exogenous variables of service quality and complaint handling. In another way, we can say the R square (R^2) of 0.621 shows the amount of the effect of exogenous variables of service quality and complaint handling on the customer satisfaction endogenous variable.

This value can be made in the form of percent numbers by calculating Coefficient of Determination (CD) by using the following formula.

CD = R^2 x 100%

CD = 0.621 x 100%

CD = 62.1%

While the rest can be calculated using the following formula:

e = 1 - R^2

e = 1 − 0.621

e = 0.379

The number 0.379 (37.9%) is the influence of other factors in the model beyond the two exogenous variables studied. In other words, the variation of customer satisfaction level which can be explained by using exogenous variables of service quality and complaint handling is equal to 62.1%; while the 37.9% is caused by other variables outside of this research.

Simultaneous Hypothesis Testing to See The Effect of Service Quality and Complaint Handling Variables on a Customer Satisfaction Level Variable

This hypothesis testing is used to see the simultaneous effect of the service quality and complaint handling on customer satisfaction level. The testing uses the value of significance (sig) from the ANOVA output table. To see whether there is a linear relationship between service quality and complaint handling with customer satisfaction level we can perform the following analysis steps:

First: Determine the hypothesis to be tested
H0: There is no linear relationship between service quality and complaint handling with customer satisfaction level
H1: There is linear relationship between service quality and complaint handling with customer satisfaction level

Second: Calculate the value of significance (sig) or pvalue
The sig value shown in column Sig is 0.000

Third: use the following criteria
If the sig value < 0.05 reject H0 and accept H1
If the sig value > 0.05 accept H0 and reject H1

Fourth: take the decision of the hypothesis testing

Since the sig value is 0.000 < 0.05; therefore reject H0 and accept H1. Thus there is linear relationship between service quality and complaint handling with customer satisfaction level. Accordingly service quality and complaint handling affect customer satisfaction level significantly.

The Partial Effect of Service Quality and Complaint Handling Exogenous Variables on A Customer Satisfaction Level Endogenous

PATH ANALYSIS: DATA ANALYSIS APPLICATION

Variable

The magnitude of the effect of exogenous variables of service quality and complaint handling individually / partially on a customer satisfaction level endogenous variable can be seen from the Beta or Standardized Coefficients value. Whereas the testing of hypothesis uses t value. The values can be seen in the **Coefficient** table output. Therefore, in the following sections we will discuss the effect of exogenous variables of service quality and complaint handling on a customer satisfaction endogenous variable partially.

Relationship between a Complaint Handling Exogenous Variable and a Customer Satisfaction Level Endogenous Variable

To see whether there is a linear relationship between a complaint handling exogenous variable and a customer satisfaction endogenous variable we can perform the following analysis steps:

First: Determine the hypothesis to be tested
H0: There is no linear relationship between a complaint handling exogenous variable and a customer satisfaction endogenous variable
H1: There is linear relationship between a complaint handling exogenous variable and a customer satisfaction endogenous variable
Second: Calculate the value of observation t (t_o)
The t_o value shown in column t in the above **Coefficients** table is 6.283

Third: Calculate the number of t table (t_α) as it has been discussed in the previous chapter

From the provisions it is obtained t_α as much as 1.960

Fourth: Define the decision-making criteria that has been discussed in the previous chapter

Fifth: Make a decision of hypothesis testing results
The calculation result with IBM SPSS shows the t_o as much as 6.283 is bigger than t_α as much as 1.960; thus the decision is reject H0 and accept H1. This means that there is a linear relationship between a complaint

handling exogenous variable and a customer satisfaction level endogenous variable. Because there is a linear relationship between the two variables; then the complaint handling affects the customer satisfaction level significantly. The amount of effect can be known from coefficient value of Beta (in column **Standardized Coefficient Beta**) as much as 0.491. The effect of this magnitude is significant because the value of significance in the **Sig column** is 0.000 which is smaller than 0.05.

Relationship between a Service Quality Exogenous Variable and a Customer Satisfaction Level Endogenous Variable

To see whether there is a linear relationship between a service quality exogenous variable and a customer satisfaction endogenous variable we can perform the following analysis steps:

First: Determine the hypothesis to be tested
H0: There is no linear relationship between a service quality exogenous variable and a customer satisfaction endogenous variable
H1: There is linear relationship between a service quality exogenous variable and a customer satisfaction endogenous variable

Second: Calculate the value of observation t (t_o)
The t_o value shown in column t in the above **Coefficients** table is 4.981
Third: Calculate the number of t table (t_α) as it has been discussed in the previous chapter

From the provisions it is obtained t_α as much as 1.960

Fourth: Define the decision-making criteria that has been discussed in the previous chapter

Fifth: Make a decision of hypothesis testing results
The calculation result with IBM SPSS shows the t_o as much as 4.981 is bigger than t_α as much as 1.960; thus the decision is reject H0 and accept H1. This means that there is a linear relationship between a service quality exogenous variable and a customer satisfaction level endogenous variable. Because there is a linear relationship between the two variables; then the service quality affects the customer satisfaction level significantly. The

amount of effect can be known from coefficient value of Beta (in column **Standardized Coefficient Beta**) as much as 0.389. The effect of this magnitude is significant because the value of significance in the **Sig column** is 0.000 which is smaller than 0.05.

Sub-Structure 3

To calculate the sub – structure 3, the steps are as follows:

- Click **Analyze> Linear**
- Enter the customer loyalty variable to the **Dependent** column
- Enter the customer satisfaction variable to **Independent** column
- **Statistics:** at the option of **Regression Coefficients**, check **Estimate, Model Fit** and **Descriptives > Continue**
- Click **Ok** to process

The result of the calculation is as follows:

Model Summary

Model	R	R Square	Adjusted R Square	Std. Error of the Estimate
1	,702ª	,493	,488	1,870

a. Predictors: (Constant), Customer Satisfaction

ANOVAª

Model		Sum of Squares	Df	Mean Square	F	Sig.
1	Regression	333,201	1	333,201	95,267	,000ᵇ
	Residual	342,759	98	3,498		
	Total	675,960	99			

a. Dependent Variable: Customer Loyalty
b. Predictors: (Constant), Customer Satisfaction

Model		Unstandardized Coefficients		Standardized Coefficients	t	Sig.
		B	Std. Error	Beta		
1	(Constant)	2,536	,990		2,562	,012
	Customer Satisfaction	,621	,064	,702	9,761	,000

a. Dependent Variable: Customer Loyalty

The Interpretation of the Calculation Result for the Sub Structure 3

The Effect of an Customer Satisfaction Variable on a Customer Loyalty Variable

The effect of the customer satisfaction exogenous variable on the customer loyalty variable can be seen in the output in the **Model Summary** table, in the R square value. The value of R square (R^2) in the table above is 0.493. This number shows the amount of variation of the customer loyalty endogenous variable that can be explained by an exogenous variable of customer satisfaction. In another way, we can say the R square (R^2) of 0.493 shows the amount of the effect of the exogenous variable of customer satisfaction on the customer loyalty endogenous variable.

This value can be made in the form of percent numbers by calculating Coefficient of Determination (CD) by using the following formula.

CD = R^2 x 100%

CD = 0.493 x 100%

CD = 49.3%

While the rest can be calculated using the following formula:

e = 1 - R^2

e = 1 − 0.493

e = 0.507

The number 0.507 (50.7%) is the influence of other factors in the model beyond the customer satisfaction exogenous variable studied. In other words, the variation of customer loyalty which can be explained by using the exogenous variable of customer satisfaction is equal to 49.3%; while the 50.7% is caused by other variables outside of this research

Hypothesis Testing to See The Effect of a Exogenous Customer Satisfaction Level Variable on a Endogenous Customer Loyalty Variable

This hypothesis testing is used to see the simultaneous effect of the customer satisfaction level on customer loyalty. The testing uses the value of significance (sig) from the ANOVA output table. To see whether there is a linear relationship between customer satisfaction level and customer loyalty we can perform the following analysis steps:

First: Determine the hypothesis to be tested
H0: There is no linear relationship between customer satisfaction level and customer loyalty
H1: There is linear relationship between customer satisfaction level and customer loyalty

Second: Calculate the value of significance (sig) or pvalue
The sig value shown in column Sig is 0.000

Third: use the following criteria
If the sig value < 0.05 reject H0 and accept H1
If the sig value > 0.05 accept H0 and reject H1

Fourth: take the decision of the hypothesis testing

Since the sig value is 0.000 < 0.05; therefore reject H0 and accept H1. Thus there is linear relationship between customer satisfaction level and customer loyalty. Accordingly customer satisfaction level affect customer loyalty significantly.

Relationship between a Customer Satisfaction Exogenous Variable and a Customer Loyalty Endogenous Variable

To see whether there is a linear relationship between a customer satisfaction exogenous variable and customer loyalty endogenous variable we can perform the following analysis steps:

First: Determine the hypothesis to be tested
H0: There is no linear relationship between a customer satisfaction exogenous variable and a customer loyalty variable
H1: There is linear relationship between a customer satisfaction exogenous variable and a customer loyalty endogenous variable

Second: Calculate the value of observation t (t_o)
The t_o value shown in column t in the above **Coefficients** table is 9.761

Third: Calculate the number of t table (t_α) as it has been discussed in the previous chapter
From the provisions it is obtained t_α as much as 1.960

Fourth: Define the decision-making criteria as it has been discussed in the previous chapter

Fifth: Make a decision of hypothesis testing results
The calculation result with IBM SPSS shows the t_o as much as 9.761 is bigger than t_α as much as 1.960; thus the decision is reject H0 and accept H1. This means that there is a linear relationship between a customer satisfaction exogenous variable and a customer loyalty endogenous variable. Because there is a linear relationship between the two variables; then the customer satisfaction affects the customer loyalty significantly. The amount of effect can be known from coefficient value of Beta (in column **Standardized Coefficient Beta**) as much as 0.702. The effect of this magnitude is significant because the value of significance in the **Sig column** is 0.000 which is smaller than 0.05

Sub-Structure 4
To calculate the sub – structure 4, the steps are as follows:

PATH ANALYSIS: DATA ANALYSIS APPLICATION

- Click **Analyze> Linear**
- Enter the customer loyalty variable to the **Dependent** column
- Enter the complaint handling and service quality variables to **Independent** column
- **Statistics:** at the option of **Regression Coefficients**, check **Estimate, Model Fit** and **Descriptives > Continue**
- Click **Ok** to process

The result of the calculation is as follows:

Model Summary

Model	R	R Square	Adjusted R Square	Std. Error of the Estimate
1	,743ª	,552	,543	1,767

a. Predictors: (Constant), Service Quality, Complaint Handling

ANOVAª

Model		Sum of Squares	Df	Mean Square	F	Sig.
1	Regression	372,996	2	186,498	59,711	,000ᵇ
	Residual	302,964	97	3,123		
	Total	675,960	99			

a. Dependent Variable: Customer Loyalty
b. Predictors: (Constant), Service Quality, Complaint Handling

Coefficientsª

Model		Unstandardized Coefficients		Standardized Coefficients	t	Sig.
		B	Std. Error	Beta		
1	(Constant)	2,176	,978		2,226	,028
	Complaint Handling	,522	,071	,626	7,371	,000
	Service Quality	,155	,076	,172	2,028	,045

a. Dependent Variable: Customer Loyalty

The Interpretation of the Calculation Result for the Sub Structure 4

The Effect of Combined Service Quality and Complaint Handling Exogenous Variables on Customer Loyalty

The effect of service quality and complaint handling exogenous variables on the customer loyalty variable can be seen in the output in the **Model Summary** table, in the R square value. The value of R square (R^2) in the table above is 0.552. This number shows the amount of variation of the customer loyalty endogenous variable that can be explained by exogenous variables of service quality and complaints handling. In another way, we can say the R square (R^2) of 0.552 shows the amount of the effect of exogenous variables of service quality and complaints handling on the customer loyalty endogenous variable.

This value can be made in the form of percent numbers by calculating Coefficient of Determination (CD) by using the following formula.

$CD = R^2 \times 100\%$

$CD = 0.552 \times 100\%$

$CD = 55.2\%$

While the rest can be calculated using the following formula:

$e = 1 - R^2$

$e = 1 - 0.552$

$e = 0.448$

The number 0.448 (44.8%) is the influence of other factors in the model beyond the two exogenous variables studied. In other words, the variation of customer loyalty level which can be explained by using exogenous variables of service quality and complaint handling is equal to 55.2%; while the 44.8% is caused by other variables outside of this research.

Simultaneous Hypothesis Testing to See The Effect of Service Quality and Complaint Handling Variables on a Customer Loyalty Variable

This hypothesis testing is used to see the simultaneous effect of the service quality and complaint handling on customer loyalty. The testing uses the value of significance (sig) from the ANOVA output table. To see whether there is a linear relationship between service quality and complaint handling with customer loyalty we can perform the following analysis steps:

First: Determine the hypothesis to be tested

H0: There is no linear relationship between service quality and complaint handling with customer loyalty

H1: There is linear relationship between service quality and complaint handling with customer loyalty

Second: Calculate the value of significance (sig) or pvalue
The sig value shown in column Sig is 0.000

Third: use the following criteria
If the sig value < 0.05 reject H0 and accept H1
If the sig value > 0.05 accept H0 and reject H1

Fourth: take the decision of the hypothesis testing

Since the sig value is 0.000 < 0.05; therefore reject H0 and accept H1. Thus there is linear relationship between service quality and complaint handling with customer loyalty. Accordingly service quality and complaint handling affect loyalty significantly.

The Partial Effect of Service Quality and Complaint Handling Exogenous Variables on A Customer Loyalty Endogenous Variable

The magnitude of the effect of exogenous variables of service quality and complaint handling individually / partially on a customer loyalty endogenous variable can be seen from the Beta or Standardized Coefficients value. Whereas the testing of hypothesis uses t value. The values can be seen in the **Coefficient** table output. Therefore, in the following sections we will discuss the effect of exogenous variables of service quality and complaint handling on a customer loyalty endogenous variable partially.

Relationship between a Complaint Handling Exogenous Variable

and a Customer Loyalty Endogenous Variable

To see whether there is a linear relationship between a complaint handling exogenous variable and a customer loyalty endogenous variable we can perform the following analysis steps:

First: Determine the hypothesis to be tested
H0: There is no linear relationship between a complaint handling exogenous variable and a customer loyalty endogenous variable
H1: There is linear relationship between a complaint handling exogenous variable and a customer loyalty endogenous variable

Second: Calculate the value of observation t (t_o)
The t_o value shown in column t in the above **Coefficients** table is 7.371

Third: Calculate the number of t table (t_α) as it has been discussed in the previous chapter

From the provisions it is obtained t_α as much as 1.960

Fourth: Define the decision-making criteria that has been discussed in the previous chapter

Fifth: Make a decision of hypothesis testing results
The calculation result with IBM SPSS shows the t_o as much as 7.371 is bigger than t_α as much as 1.960; thus the decision is reject H0 and accept H1. This means that there is a linear relationship between a complaint handling exogenous variable and a customer loyalty endogenous variable. Because there is a linear relationship between the two variables; then the complaint handling affects the customer loyalty significantly. The amount of effect can be known from coefficient value of Beta (in column **Standardized Coefficient Beta**) as much as 0.626. The effect of this magnitude is significant because the value of significance in the **Sig column** is 0.000 which is smaller than 0.05.

Relationship between a Service Quality Exogenous Variable and a Customer Loyalty Endogenous Variable

To see whether there is a linear relationship between a service quality

PATH ANALYSIS: DATA ANALYSIS APPLICATION

exogenous variable and a customer loyalty endogenous variable we can perform the following analysis steps:

First: Determine the hypothesis to be tested
H0: There is no linear relationship between a service quality exogenous variable and a customer loyalty endogenous variable
H1: There is linear relationship between a service quality exogenous variable and a customer loyalty endogenous variable

Second: Calculate the value of observation t (t_o)
The t_o value shown in column t in the above **Coefficients** table is 2.028

Third: Calculate the number of t table (t_α) as it has been discussed in the previous chapter
From the provisions it is obtained t_α as much as 1.960

Fourth: Define the decision-making criteria that has been discussed in the previous chapter

Fifth: Make a decision of hypothesis testing results
The calculation result with IBM SPSS shows the t_o as much as 2.028 is bigger than t_α as much as 1.960; thus the decision is reject H0 and accept H1. This means that there is a linear relationship between a service quality exogenous variable and a customer loyalty endogenous variable. Because there is a linear relationship between the two variables; then the service quality affects the customer loyalty significantly. The amount of effect can be known from coefficient value of Beta (in column **Standardized Coefficient Beta**) as much as 0.172. The effect of this magnitude is significant because the value of significance in the **Sig column** is 0.045 which is smaller than 0.05.

Calculating Direct Effect (DE)

To calculate the direct (DE) effect is as follows:
- The influence of complaint handling to service quality X1 to X2 = 0.600

- The influence of complaint handling to customer satisfaction X1 to Y1 = 0.491
- The influence of service quality to customer satisfaction X2 to Y1 = 0.389
- The influence of complaint handling to customer loyalty X1 to Y2 = 0.457
- The influence of service quality to customer loyalty X2 to Y2 = 0.038
- The influence of customer satisfaction on customer loyalty Y1 to Y2 = 0.345

Calculating Indirect Effect (IE)

To calculate the indirect effect (IE) is as follows:

- The influence of complaint handling to customer loyalty through customer satisfaction
  PY1X1 x PY2Y1 = (0.491 x 0.345) = 0.169
- The influence of service quality on customer loyalty through customer satisfaction
  PY1X2 x PY2Y1 = (0.389 x 0.345) = 0.134

Calculating the Total Effect (TE)

To calculate the total effect (IE) is as follows:

- The influence of complaint handling to customer loyalty through customer satisfaction
  PY1X1 + PY2Y1 = (0.491 + 0.345) = 0.836
- The influence of service quality on customer loyalty through customer satisfaction
  PY1X2 + PY2Y1 = (0.389 + 0.345) = 0.374

Creating a Path Diagram of a Joint (Combined) Multiple Regression and Mediation Model

The following is the result of calculation of the parameters in the path diagram of the above model.

PATH ANALYSIS: DATA ANALYSIS APPLICATION

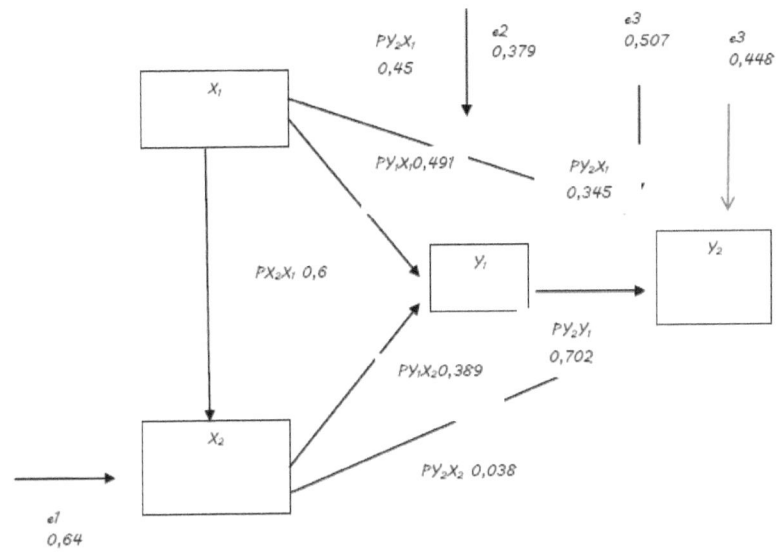

Figure 7.4 A Path Diagram and the Values of The Parameter for the Joint Model

The path diagram has the following structural equations:
Substructure 1: X2 = 0.60 X1 + e1
Substructure 2: Y1 = 0.491 X1 + 0.389X2 + e2
Substructure 3: Y2 = 0.345 Y1 + e3
Substructure 4: Y2: 0,457X1 + 0,038X2 + e4

Making Conclusions for the Results of the whole model

Based on the results of the calculation analysis above, we can draw the following conclusions:

- The effect of complaint handling on service quality is 0.6
- The combining effect of complaint handling and service quality on customer satisfaction is 0.621
- The effect of complaint handling on customer satisfaction is 0.491
- The effect of service quality on customer satisfaction is 0.389

- The effect of customer satisfaction on customer loyalty is 0.702
- The combining effect of complaint handling and service quality on customer loyalty is 0.552
- The effect of complaint handling on customer loyalty is 0.345
- The effect of service quality on customer loyalty is 0.038

CHAPTER 8

AN APPLICATION OF A COMPLEX MODEL

8.1 A Case Example

In this example of the complex model we will use three independent variables that serve as exogenous variables and three dependent variables that serve as endogenous variables. As the exogenous variables are the complaint handling, the service support facility and the service process variables, while the first endogenous variable serving as the intervening variable is the quality of the service; and the second one is customer satisfaction and the third endogenous variable is customer loyalty. The path diagram model is based on theory is as follows:

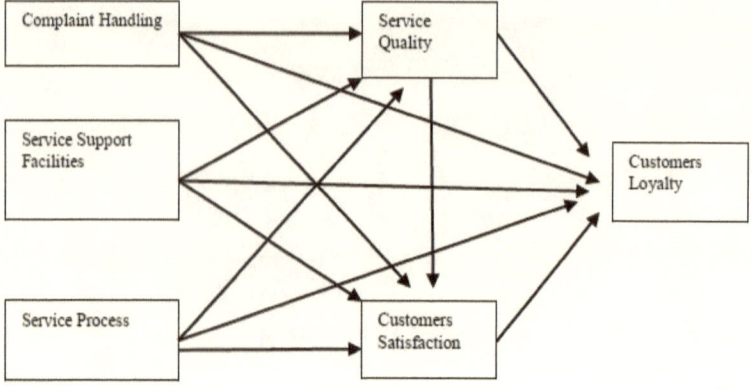

Figure 8.1 A Complex Path Diagram Model of Complaint Handling, Service Support, Service Process, Service Quality, Customer Satisfaction and Customer Loyalty Variables ' Relationship Based on Theory

The model can be broken down into 3 parts, as follows:

- First: Complaint handling, service support facilities, and service processes variables affect both partially and simultaneously on service quality
- Second: Complaint handling, service support facilities, service processes, and service quality variables affect both partially and simultaneously on customer satisfaction
- Third: Complaint handling, service support facilities, service process, service quality and customer satisfaction affect both partially and simultaneously on customer loyalty

8.2 Hypotheses and Problems

The first hypothesis is as follows:
H0: Complaint handling , service support facilities, and service processes variables do not affect both partially and simultaneously on service quality
H1: Complaint handling , service support facilities, and service processes variables affect both partially and simultaneously on service quality

PATH ANALYSIS: DATA ANALYSIS APPLICATION

The second hypothesis is as follows:
H0: Complaint handling, service support facilities, service process, and service quality do not affect both partially and simultaneously on customer satisfaction
H1: Complaint handling, service support facilities, service process, and service quality affect both partially and simultaneously on customer satisfaction

The third hypothesis is as follows:
H0: Complaint handling, service support facilities, service process, service quality and customer satisfaction variables do not affect both partially and simultaneously on customer loyalty
H1: Complaint handling, service support facilities, service process, service quality and customer satisfaction variables affect both partially and simultaneously on customer loyalty

The problems for this case are as follows:
a. How much is the influence of complaint handling, service support facility, and service process variables on service quality either simultaneously or partially?
b. How much is the influence of complaint handling, service support facility, service process and service quality variables on customer satisfaction either simultaneously or partially?
c. How much is the influence of complaint handling, service support facility, service process, service quality and customer satisfaction variables on customer loyalty either simultaneously or partially?
d. How much is the influence of complaint handling, service support facility, and service process through service quality on customer satisfaction?
e. How much is the influence of complaint handling, service support facility, and service process variables through service quality on customer loyalty?
f. How much is the influence of complaint handling, service support facility, service process and service quality variables through customer satisfaction on customer loyalty?

8.3 Research Data

There are 100 data for this case as follows.

No	process	facility	complaint	service	satisfaction	loyalty
1	16	17	18	17	16	15
2	12	13	15	17	18	12
3	15	16	17	14	16	14
4	12	13	14	14	14	13
5	11	12	15	15	16	12
6	14	15	17	15	16	13
7	11	12	13	16	12	14
8	15	16	19	19	20	11
9	13	14	15	16	17	14
10	15	16	19	19	18	14
11	10	11	14	16	17	12
12	12	13	15	11	15	10
13	11	12	14	14	13	12
14	14	15	16	16	18	11
15	10	11	10	14	16	12
16	11	12	13	15	17	13
17	16	17	19	12	17	14
18	13	14	15	12	16	14
19	11	12	15	12	14	15
20	13	14	16	14	15	14
21	10	11	12	11	15	14
22	14	15	18	16	14	14
23	12	13	13	15	15	11
24	12	13	14	14	13	12
25	11	12	13	13	17	11
26	10	11	10	10	12	9
27	11	12	13	17	13	9
28	10	11	12	12	11	10
29	11	12	11	14	12	8
30	10	10	9	14	8	7
31	10	11	10	10	13	12
32	10	9	8	12	12	9
33	11	12	13	14	12	10

PATH ANALYSIS: DATA ANALYSIS APPLICATION

34	11	9	8	15	14	9
35	13	12	11	8	12	10
36	12	11	13	13	16	11
37	11	10	11	14	11	9
38	11	10	9	15	14	11
39	13	14	15	11	12	12
40	12	10	11	12	15	8
41	13	10	11	12	13	8
42	12	11	9	14	14	9
43	11	12	10	17	11	9
44	11	11	9	10	12	8
45	10	11	12	9	12	7
46	11	12	13	15	15	8
47	13	14	13	8	14	10
48	12	13	14	11	11	9
49	11	12	11	13	10	9
50	10	11	12	13	15	13
51	11	12	13	14	13	12
52	13	12	13	13	14	10
53	13	14	15	16	16	15
54	12	11	12	12	14	12
55	11	12	13	12	15	9
56	13	14	15	15	16	14
57	12	13	14	18	12	10
58	11	12	13	11	13	11
59	12	11	10	11	13	12
60	14	15	16	18	16	12
61	13	14	14	11	11	10
62	14	16	17	12	12	13
63	13	15	16	14	13	12
64	12	13	14	15	12	10
65	11	12	13	11	14	10
66	11	12	13	11	14	11
67	12	13	12	11	14	9
68	10	11	13	14	13	10
69	13	14	15	12	13	9

70	11	12	13	13	17	12	
71	11	12	13	14	12	11	
72	12	11	13	16	15	12	
73	13	14	15	14	15	12	
74	12	13	14	13	15	10	
75	11	12	14	14	13	10	
76	13	14	16	16	18	15	
77	15	17	20	19	22	17	
78	14	16	17	20	20	16	
79	13	14	16	20	20	15	
80	16	17	18	16	16	16	
81	12	14	15	18	19	12	
82	14	15	16	19	19	12	
83	17	18	20	16	15	13	
84	18	20	21	22	22	19	
85	15	17	19	19	19	12	
86	14	16	18	15	16	13	
87	15	17	18	19	20	12	
88	16	18	17	15	18	12	
89	18	20	21	17	20	17	
90	14	15	16	17	19	17	
91	13	15	16	19	18	16	
92	15	17	19	16	20	16	
93	14	16	17	19	18	16	
94	14	16	19	18	21	13	
95	18	20	21	17	20	19	
96	17	18	19	16	17	15	
97	16	17	19	15	19	15	
98	17	18	17	19	19	15	
99	15	17	19	19	19	12	
100	14	17	18	17	19	14	

Stages in Completing the Case

To solve the case the stages are as below:

First: Creating a path diagram model based on the relationship between the variables we are examining as follows:

PATH ANALYSIS: DATA ANALYSIS APPLICATION

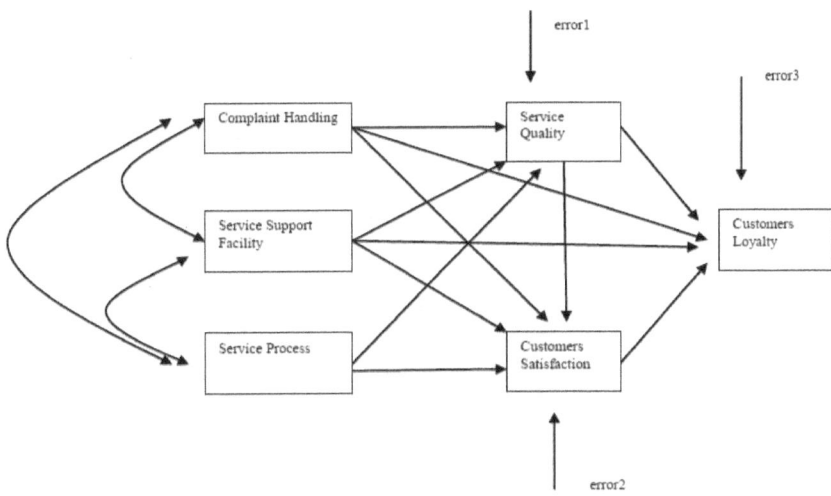

Figure 8.2 A Complex Path Diagram Model of Complaint Handling, Service Support, Service Process, Service Quality, Customer Satisfaction and Customer Loyalty Variables' Relationship

Second: Create a path diagram of the above model

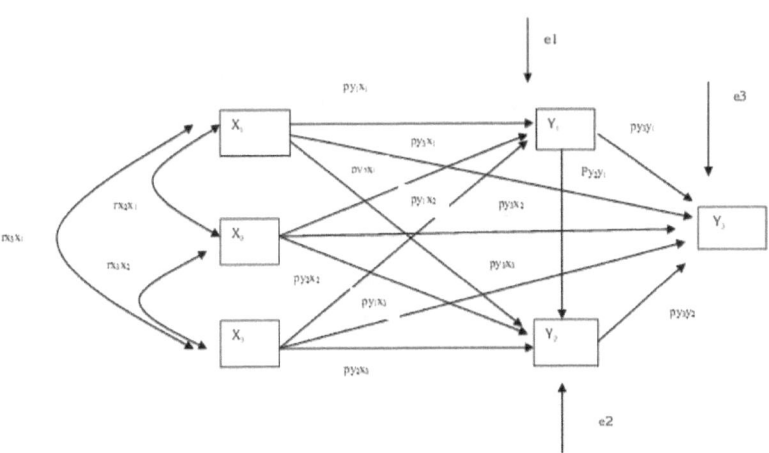

Figure 8.3 A Complex Path Diagram of Complaint Handling, Service Support, Service Process, Service Quality, Customer Satisfaction and Customer Loyalty Variables' Relationship

Where:
X1 as a complaint handling exogenous variable
X2 as a facility support exogenous variable
X3 as a service process exogenous variable
Y1 as a service quality intervening endogenous variable
Y2 as a customer satisfaction intervening endogenous variable
Y3 as a customer loyalty endogenous variable

Third: Create a structural equation

The path diagram above has three structural equations. The structural equation can be seen as below:

- $Y1 = PY1X1 + PY\ X + PY\ X3 + e1$ (As a sub-structure equation 1)
- $Y2 = PY\ X + PY\ X2 + PY\ X3 + PY\ Y + e2$ (As a sub-structure equation 2)
- $Y3 = PY3X + PY3X2 + PY3X3 + PY3Y2 + PY3Y + e3$ (As a sub-structure equation 3)

Fourth: Creating a variable design, entering data and presenting it in IBM SPSS Design variables can be created by selecting the left bottom sub menu on the command: **Variable View**. Write down all variables pursuant to the example below

Name	Type	Width	Decimal	Label	Values	Missing	Column	Align	Measure	Role
complaint	Numeric	8	0	Complaint Handling	None	None	8	Right	Scale	Input
facility	Numeric	8	0	Support facility	None	None	8	Right	Scale	Input
process	Numeric	8	0	Service process	None	None	8	Right	Scale	Input
service	Numeric	8	0	Service Quality	None	None	8	Right	Scale	Both
satisfaction	Numeric	8	0	Satisfaction level	None	None	8	Right	Scale	Both
loyalty	Numeric	8	0	Customer Loyalty	None	None	8	Right	Scale	Target

PATH ANALYSIS: DATA ANALYSIS APPLICATION

Fifth: Enter data by clicking on the command: **Data View**

No	complaint	facility	process	service	satisfaction	loyalty
1						
.						
.						
100						

Conducting Calculations for Sub - Structure 1:
Sixth: Perform analysis by using IBM SPSS with the following steps

Sub Structure 1

- Click **Analyze> Linear**
- Enter the service quality variable to the **Dependent** column
- Enter the complaint handling, service support facilities, and service process variables to **Independent** column
- **Statistics:** at the option of **Regression Coefficients**, check **Estimate, Model Fit** and **Descriptives > Continue**
- Click **Ok** to process

The result of calculation is as follows:

Correlations

		Service Quality	Service Process	Support Facilities	Complaint Handling
Pearson Correlation	Service Quality	1,000	,525	,574	,600
	Service Process	,525	1,000	,933	,854
	Support Facilities	,574	,933	1,000	,933
	Complaint Handling	,600	,854	,933	1,000
Sig. (1-tailed)	Service Quality	.	,000	,000	,000
	Service Process	,000	.	,000	,000
	Support Facilities	,000	,000	.	,000
	Complaint Handling	,000	,000	,000	.
N	Service Quality	100	100	100	100
	Service Process	100	100	100	100
	Support Facilities	100	100	100	100
	Complaint Handling	100	100	100	100

Model Summary

Model	R	R Square	Adjusted R Square	Std. Error of the Estimate
1	,601a	,362	,342	2,355

a. Predictors: (Constant), Complaint Handling, Service Process, Support Facilities

ANOVAa

Model		Sum of Squares	df	Mean Square	F	Sig.
1	Regression	301,789	3	100,596	18,133	,000b
	Residual	532,571	96	5,548		
	Total	834,360	99			

a. Dependent Variable: Service Quality
b. Predictors: (Constant), Complaint Handling, Service Process, Support Facilities

Coefficientsa

Model		Unstandardized Coefficients		Standardized Coefficients	t	Sig.
		B	Std. Error	Beta		
1	(Constant)	6,266	1,524		4,111	,000
	Service Process	-,021	,323	-,015	-,064	,949
	Support Facilities	,134	,376	,118	,356	,722
	Complaint Handling	,465	,212	,503	2,192	,031

a. Dependent Variable: Service Quality

Seventh: The interpretation of the calculation results using IBM SPSS

The Effect of Combined Exogenous Service Process, Support Facilities and Complaint Handling Variables on a Service Quality Variable

The effect of exogenous service process, support facilities and complaint handling variables on a service quality variable can be seen in the output in the **Model Summary** table, in the R square value. The value of R square (R^2) in the table above is 0.362. This number shows the amount of variation of the service quality variable that can be explained by exogenous variables of service process, support facilities and complaint handling. In another way, we can say the R square (R^2) of 0.362 shows the amount of the effect of exogenous variables of service process, support facilities and complaint handling on the service quality endogenous variable.

This value can be made in the form of percent numbers by calculating Coefficient of Determination (CD) by using the following formula.

CD = R^2 x 100%

CD = 0.362 x 100%

CD = 36.2%

While the rest can be calculated using the following formula:

e = 1 - R^2

e = 1 - 0.362

e = 0.638

The number 0.638 (63.8%) is the influence of other factors in the model beyond the three exogenous variables studied. In other words, the variation of service quality which can be explained by using exogenous variables of service process, support facilities and complaint handling is equal to 36.2%; while the 63.8% is caused by other variables outside of this research.

Simultaneous Hypothesis Testing to See The Effect of Exogenous Service Process, Support Facilities and Complaint Handling Variables on a Service Quality Variable

This hypothesis testing is used to see the simultaneous effect of the service process, support facilities and complaint handling variables on service quality. The testing uses the value of significance (sig) from the ANOVA output table. To see whether there is a linear relationship between service process, support facilities and complaint handling with service quality we can perform the following analysis steps:

First: Determine the hypothesis to be tested
H0: There is no linear relationship between service process, support facilities and complaint handling with service quality
H1: There is linear relationship between service process, support facilities and complaint handling with service quality

Second: Calculate the value of significance (sig) or pvalue
The sig value shown in column Sig is 0.000

Third: use the following criteria
If the sig value < 0.05 reject H0 and accept H1
If the sig value > 0.05 accept H0 and reject H1

Fourth: take the decision of the hypothesis testing

Since the sig value is 0.000 < 0.05; therefore reject H0 and accept H1. Thus there is linear relationship between service process, support facilities and complaint handling with service quality. Accordingly service process, support facilities and complaint handling affect service quality significantly.

The Partial Effect of Service Process, Support Facilities, and Complaint Handling Variables on A Service Quality Endogenous Variable

The magnitude of the effect of exogenous variables of service process, support facilities, and complaint handling individually / partially on a service quality endogenous variable can be seen from the Beta or Standardized Coefficients value. Whereas the testing of hypothesis uses t value. The values can be seen in the **Coefficient** table output. Therefore, in the following sections we will discuss the effect of exogenous variables of service process, support facilities, and complaint handling on a service quality endogenous variable partially.

PATH ANALYSIS: DATA ANALYSIS APPLICATION

Relationship between a Service Process Exogenous Variable and a Quality Service Endogenous Variable

To see whether there is a linear relationship between a service process exogenous variable and a service quality endogenous variable we can perform the following analysis steps:

First: Determine the hypothesis to be tested
H0: There is no linear relationship between a service process exogenous variable and a service quality endogenous variable
H1: There is linear relationship between a service process exogenous variable and a service quality endogenous variable

Second: Calculate the value of observation t (t_o)
The t_o value shown in column t in the above **Coefficients** table is -0.064

Third: Calculate the number of t table (t_α) as it has been discussed in the previous chapter:

From these provisions it is obtained t_α as much as 1.960

Fourth: Define the decision-making criteria as follows
If $t_o > t_\alpha$, then H0 is rejected and H1 accepted;
If $t_o < t_\alpha$, then H0 is accepted and H1 is rejected
or we can use the second alternative by using the significance level (probability / p value)
If sig < 0.05, then there is significant influence
If sig > 0.05, then there is no significant influence

Fifth: Make a decision of hypothesis testing results
The calculation result with IBM SPSS shows the t_o as much as -0.0641. Since the t_o is negative, to make easier we can use the significance instead of t. The value of significance for the t_o in Column **Sig** is as much as 0.949. Since the sig from the calculation as much as 0.949 is bigger than 0.05 then there is no a significant influence of the relationship between the two variables.

Relationship between a Support Facilities Exogenous Variable and a Quality Service Endogenous Variable

To see whether there is a linear relationship between a support facilities exogenous variable and a service quality endogenous variable we can perform the following analysis steps:

First: Determine the hypothesis to be tested
H0: There is no linear relationship between a support facilities exogenous variable and a service quality endogenous variable
H1: There is linear relationship between a support facilities exogenous variable and a service quality endogenous variable

Second: Calculate the value of observation t (t_o)
The t_o value shown in column t in the above **Coefficients** table is 0.356

Third: Calculate the number of t table (t_α) as it has been discussed in the previous chapter:

From these provisions it is obtained t_α as much as 1.960

Fourth: Define the decision-making criteria as follows
If $t_o > t_\alpha$, then H0 is rejected and H1 accepted;
If $t_o < t_\alpha$, then H0 is accepted and H1 is rejected
or we can use the second alternative by using the significance level (probability / p value)
If sig < 0.05, then there is significant influence
If sig > 0.05, then there is no significant influence

Fifth: Make a decision of hypothesis testing results
The calculation result with IBM SPSS shows the t_o as much as 0.356 is smaller than t_α as much as 1.960; thus the decision is accept H0 and reject H1. This means that there is no a linear relationship between a support facility exogenous variable and a service quality endogenous variable.

PATH ANALYSIS: DATA ANALYSIS APPLICATION

Relationship between a Complaint Handling Exogenous Variable and a Quality Service Endogenous Variable

To see whether there is a linear relationship between a complaint handling exogenous variable and a service quality endogenous variable we can perform the following analysis steps:

First: Determine the hypothesis to be tested
H0: There is no linear relationship between a complaint handling exogenous variable and a service quality endogenous variable
H1: There is linear relationship between a complaint handling exogenous variable and a service quality endogenous variable

Second: Calculate the value of observation t (t_o)
The t_o value shown in column t in the above **Coefficients** table is 2.192

Third: Calculate the number of t table (t_α) as it has been discussed in the previous chapter:

From these provisions it is obtained t_α as much as 1.960

Fourth: Define the decision-making criteria as it has been discussed above.

Fifth: Make a decision of hypothesis testing results

The calculation result with IBM SPSS shows the t_o as much as 2.192 is bigger than t_α as much as 1.960; thus the decision is reject H0 and accept H1. This means that there is a linear relationship between a complaint handling exogenous variable and a service quality endogenous variable. Because there is a linear relationship between the two variables; then the complaint handling affects the service quality significantly. The amount of effect can be known from coefficient value of Beta (in column **Standardized Coefficient Beta**) as much as 0.503. The effect of this magnitude is significant because the value of significance in the **Sig column** is 0.031 which is smaller than 0.05.

Calculation for Sub Structure II

- Click **Analyze> Linear**
- Enter the customer satisfaction variable to the **Dependent** column
- Enter the complaint handling, service support facilities, service process and service quality variables to **Independent** column
- **Statistics:** at the option of **Regression Coefficients**, check **Estimate, Model Fit** and **Descriptives > Continue**
- Click **Ok** to process

The result of calculation is as follows:

Correlations

		Customer Satisfaction	Service Process	Support Facilities	Complaint Handling	Service Quality
Pearson Correlation	Customer Satisfaction	1,000	,651	,688	,724	,684
	Service Process	,651	1,000	,933	,854	,525
	Support Facilities	,688	,933	1,000	,933	,574
	Complaint Handling	,724	,854	,933	1,000	,600
	Service Quality	,684	,525	,574	,600	1,000
Sig. (1-tailed)	Customer Satisfaction		,000	,000	,000	,000
	Service Process	,000		,000	,000	,000
	Support Facilities	,000	,000		,000	,000
	Complaint Handling	,000	,000	,000		,000
	Service Quality	,000	,000	,000	,000	
N	Customer Satisfaction	100	100	100	100	100
	Service Process	100	100	100	100	100
	Support Facilities	100	100	100	100	100
	Complaint Handling	100	100	100	100	100
	Service Quality	100	100	100	100	100

Model Summary

Model	R	R Square	Adjusted R Square	Std. Error of the Estimate
1	,791a	,625	,609	1,846

a. Predictors: (Constant), Service Quality, Service Process, Complaint Handling, Support Facilities

ANOVAa

Model		Sum of Squares	df	Mean Square	F	Sig.
1	Regression	539,917	4	134,979	39,602	,000b
	Residual	323,793	95	3,408		
	Total	863,710	99			

a. Dependent Variable: Customer Satisfaction
b. Predictors: (Constant), Service Quality, Service Process, Complaint Handling, Support Facilities

Coefficientsa

Model		Unstandardized Coefficients		Standardized Coefficients	t	Sig.
		B	Std. Error	Beta		
1	(Constant)	2,053	1,295		1,585	,116
	Service Process	,232	,253	,161	,915	,362
	Support Facilities	-,132	,295	-,114	-,447	,656
	Complaint Handling	,433	,171	,460	2,542	,013
	Service Quality	,395	,080	,388	4,939	,000

a. Dependent Variable: Customer Satisfaction

Second: The interpretation of the calculation results using IBM SPSS The Effect of Combined Exogenous Service Process, Support Facilities, Complaint Handling and Service Quality Variables on a Customer Satisfaction Variable

The effect of exogenous service process, support facilities, complaint handling and service quality variables on a customer satisfaction variable can be seen in the output in the **Model Summary** table, in the R square value. The value of R square (R^2) in the table above is 0.625. This number shows the amount of variation of the customer satisfaction variable that can be explained by exogenous variables of service process, support facilities, complaint handling and service quality. In another way, we can say the R square (R^2) of 0.625 shows the amount of the effect of exogenous variables of service process, support facilities, complaint handling and service quality on a customer satisfaction endogenous variable.

This value can be made in the form of percent numbers by calculating Coefficient of Determination (CD) by using the following formula.

CD = R^2 x 100%

CD = 0.625 x 100%

CD = 62.5%

While the rest can be calculated using the following formula:

e = 1 - R^2

e = 1 – 0.625

e = 0.375

The number 0.375 (37.5%) is the influence of other factors in the model beyond the four exogenous variables studied. In other words, the variation of customer satisfaction which can be explained by using exogenous variables of service process, support facilities, complaint handling and service quality is equal to 62.5%; while the 37.5% is caused by other variables outside of this research.

PATH ANALYSIS: DATA ANALYSIS APPLICATION

Simultaneous Hypothesis Testing to See The Effect of Exogenous Service Process, Support Facilities Complaint Handling and Service Quality Variables on a Customer Satisfaction Level Variable

This hypothesis testing is used to see the simultaneous effect of the service process, support facilities complaint handling and service quality on customer satisfaction level. The testing uses the value of significance (sig) from the ANOVA output table. To see whether there is a linear relationship between service process, support facilities complaint handling and service quality with customer satisfaction level we can perform the following analysis steps:

First: Determine the hypothesis to be tested
H0: There is no linear relationship between service process, support facilities complaint handling and service quality with customer satisfaction level
H1: There is linear relationship between service process, support facilities complaint handling and service quality with customer satisfaction level

Second: Calculate the value of significance (sig) or pvalue
The sig value shown in column Sig is 0.000
Third: use the following criteria
If the sig value < 0.05 reject H0 and accept H1
If the sig value > 0.05 accept H0 and reject H1

Fourth: take the decision of the hypothesis testing

Since the sig value is 0.000 < 0.05; therefore reject H0 and accept H1. Thus there is linear relationship between service process, support facilities complaint handling and service quality with customer satisfaction level. Accordingly service process, support facilities complaint handling and service quality affect customer satisfaction level significantly.

The Partial Effect of Service Process, Support Facilities, Complaint Handling and Service Quality Variables on A Customer Satisfaction Endogenous Variable

The magnitude of the effect of exogenous variables of service process, support facilities, complaint handling and service quality individually / partially on a customer satisfaction endogenous variable can be seen from the Beta or Standardized Coefficients value. Whereas the testing of hypothesis uses t value. The values can be seen in the **Coefficient** table output. Therefore, in the following sections we will discuss the effect of exogenous variables of service process, support facilities, complaint handling and service quality on a customer satisfaction endogenous variable partially.

Relationship between a Service Process Exogenous Variable and a Customer Satisfaction Endogenous Variable

To see whether there is a linear relationship between a service process exogenous variable and a customer satisfaction endogenous variable we can perform the following analysis steps:

First: Determine the hypothesis to be tested
H0: There is no linear relationship between a service process exogenous variable and a customer satisfaction endogenous variable
H1: There is linear relationship between a service process exogenous variable and a customer satisfaction endogenous variable
Second: Calculate the value of observation t (t_o)
The t_o value shown in column t in the above **Coefficients** table is 0.915

Third: Calculate the number of t table (t_α) as it has been discussed in the previous chapter:

From these provisions it is obtained t_α as much as 1.960

Fourth: Define the decision-making criteria as it has been discussed above
Fifth: Make a decision of hypothesis testing results
The calculation result with IBM SPSS shows the t_o as much as 0.915 is smaller than t_α as much as 1.960; thus the decision is accept H0 and reject H1. This means that there is no a linear relationship between a service process exogenous variable and a customer service endogenous variable.

PATH ANALYSIS: DATA ANALYSIS APPLICATION

Relationship between a Support Facilities Exogenous Variable and a Customer Satisfaction Endogenous Variable

To see whether there is a linear relationship between a support facilities exogenous variable and a customer satisfaction endogenous variable we can perform the following analysis steps:

First: Determine the hypothesis to be tested
H0: There is no linear relationship between a support facilities exogenous variable and a customer satisfaction endogenous variable
H1: There is linear relationship between a support facilities exogenous variable and a customer satisfaction endogenous variable

Second: Calculate the value of observation t (t_o)
The t_o value shown in column t in the above **Coefficients** table is - 0.447

Third: Calculate the number of t table (t_α) as it has been discussed in the previous chapter:

From these provisions it is obtained t_α as much as 1.960

Fourth: Define the decision-making criteria as it has been discussed above

Fifth: Make a decision of hypothesis testing results

The calculation result with IBM SPSS shows the t_o as much as -0.447. Since the t_o is negative, to make easier we can use the significance instead of t. The value of significance for the t_o in Column **Sig** is as much as 0.656. Since the sig from the calculation as much as 0.656 is bigger than 0.05 then there is no a significant influence of the relationship between the two variables.

Relationship between a Complaint Handling Exogenous Variable and a Customer Satisfaction Endogenous Variable

To see whether there is a linear relationship between a complaint handling exogenous variable and a customer satisfaction endogenous variable we can perform the following analysis steps:

First: Determine the hypothesis to be tested
H0: There is no linear relationship between a complaint handling exogenous variable and a customer satisfaction endogenous variable
H1: There is linear relationship between a complaint handling exogenous variable and a customer satisfaction endogenous variable

Second: Calculate the value of observation t (t_o)
The t_o value shown in column t in the above **Coefficients** table is 2.542

Third: Calculate the number of t table (t_α) as it has been discussed in the previous chapter:

From these provisions it is obtained t_α as much as 1.960

Fourth: Define the decision-making criteria as it has been discussed above

Fifth: Make a decision of hypothesis testing results

The calculation result with IBM SPSS shows the t_o as much as 2.542 is bigger than t_α as much as 1.960; thus the decision is reject H0 and accept H1. This means that there is a linear relationship between a complaint handling exogenous variable and a customer satisfaction endogenous variable. Because there is a linear relationship between the two variables; then the complaint handling affects the customer satisfaction significantly. The amount of effect can be known from coefficient value of Beta (in column **Standardized Coefficient Beta**) as much as 0.503. The effect of this magnitude is significant because the value of significance in the **Sig column** is 0.031 which is smaller than 0.05.

Relationship between a Service Quality Exogenous Variable and a Customer Satisfaction Endogenous Variable

To see whether there is a linear relationship between a service quality exogenous variable and a customer satisfaction endogenous variable we can perform the following analysis steps:

First: Determine the hypothesis to be tested
H0: There is no linear relationship between a service quality exogenous variable and a customer satisfaction endogenous variable

H1: There is linear relationship between a service quality exogenous variable and a customer satisfaction endogenous variable

Second: Calculate the value of observation t (t_o)
The t_o value shown in column t in the above **Coefficients** table is 4.939

Third: Calculate the number of t table (t_α) as it has been discussed in the previous chapter:

From these provisions it is obtained t_α as much as 1.960

Fourth: Define the decision-making criteria as it has been discussed above

Fifth: Make a decision of hypothesis testing results

The calculation result with IBM SPSS shows the t_o as much as 4.939 is bigger than t_α as much as 1.960; thus the decision is reject H0 and accept H1. This means that there is a linear relationship between a service quality exogenous variable and a customer satisfaction endogenous variable. Because there is a linear relationship between the two variables; then the service quality affects the customer satisfaction significantly. The amount of effect can be known from coefficient value of Beta (in column **Standardized Coefficient Beta**) as much as 0.503. The effect of this magnitude is significant because the value of significance in the **Sig column** is 0.031 which is smaller than 0.05.

12.8 Calculation for Sub Structure III

- Click **Analyze> Linear**
- Enter the customer loyalty variable to the **Dependent** column
- Enter the complaint handling, service support facilities, service process, service quality and customer satisfaction variables to **Independent** column
- **Statistics:** at the option of **Regression Coefficients**, check **Estimate, Model Fit** and **Descriptives > Continue**
- Click **Ok** to process

The result of calculation is as follows:

Correlations

		Customer Loyalty	Service Process	Support Facilities	Complaint Handling	Service Quality	Customer Satisfaction
Pearson Correlation	Customer Loyalty	1,000	,653	,710	,730	,548	,702
	Service Process	,653	1,000	,933	,854	,525	,651
	Support Facilities	,710	,933	1,000	,933	,574	,688
	Complaint Handling	,730	,854	,933	1,000	,600	,724
	Service Quality	,548	,525	,574	,600	1,000	,684
	Customer Satisfaction	,702	,651	,688	,724	,684	1,000
Sig. (1-tailed)	Customer Loyalty		,000	,000	,000	,000	,000
	Service Process	,000		,000	,000	,000	,000
	Support Facilities	,000	,000		,000	,000	,000
	Complaint Handling	,000	,000	,000		,000	,000
	Service Quality	,000	,000	,000	,000		,000
	Customer Satisfaction	,000	,000	,000	,000	,000	
N	Customer Loyalty	100	100	100	100	100	100
	Service Process	100	100	100	100	100	100
	Support Facilities	100	100	100	100	100	100
	Complaint Handling	100	100	100	100	100	100
	Service Quality	100	100	100	100	100	100
	Customer Satisfaction	100	100	100	100	100	100

Model Summary

Model	R	R Square	Adjusted R Square	Std. Error of the Estimate
1	,776a	,602	,580	1,693

a. Predictors: (Constant), Customer Satisfaction, Service Process, Service Quality, Complaint Handling, Support Facilities

ANOVAa

Model		Sum of Squares	Df	Mean Square	F	Sig.
1	Regression	406,605	5	81,321	28,379	,000b
	Residual	269,355	94	2,865		
	Total	675,960	99			

a. Dependent Variable: Customer Loyalty
b. Predictors: (Constant), Customer Satisfaction, Service Process, Service Quality, Complaint Handling, Support Facilities

PATH ANALYSIS: DATA ANALYSIS APPLICATION

Coefficients^a

Model		Unstandardized Coefficients		Standardized Coefficients	T	Sig.
		B	Std. Error	Beta		
1	(Constant)	1,043	1,203		,867	,388
	Service Process	-,068	,233	-,053	-,290	,772
	Support Facilities	,245	,270	,240	,908	,366
	Complaint Handling	,234	,162	,281	1,449	,151
	Service Quality	,031	,082	,034	,377	,707
	Customer Satisfaction	,304	,094	,344	3,236	,002

a. Dependent Variable: Customer Loyalty

Second: The interpretation of the calculation results using IBM SPSS

The Combined Effect of Exogenous Service Process, Support Facilities, Complaint Handling, Service Quality and Customer Satisfaction Variables on a Customer Loyalty Variable

The effect of exogenous service process, support facilities, complaint handling, service quality and customer satisfaction variables on a customer loyalty variable can be seen in the output in the **Model Summary** table, in the R square value. The value of R square (R^2) in the table above is 0.602. This number shows the amount of variation of the customer loyalty variable that can be explained by exogenous variables of service process, support facilities, complaint handling, service quality and customer satisfaction. In another way, we can say the R square (R^2) of 0.625 shows the amount of the effect of exogenous variables of service process, support facilities, complaint handling, service quality and customer satisfaction on a customer loyalty endogenous variable.

This value can be made in the form of percent numbers by calculating Coefficient of Determination (CD) by using the following formula.

CD = R^2 x 100%

CD = 0.625 x 100%

CD = 62.5%

While the rest can be calculated using the following formula:

$e = 1 - R^2$

$e = 1 - 0.625$

$e = 0.375$

The number 0.375 (37.5%) is the influence of other factors in the model beyond the four exogenous variables studied. In other words, the variation of customer loyalty which can be explained by using exogenous variables of service process, support facilities, complaint handling, service quality and customer satisfaction is equal to 62.5%; while the 37.5% is caused by other variables outside of this research.

Simultaneous Hypothesis Testing to See The Effect of Exogenous Service Process, Support Facilities Complaint Handling, Service Quality and Customer Satisfaction Level Variables on a Customer Loyalty Variable

This hypothesis testing is used to see the simultaneous effect of the service process, support facilities complaint handling, service quality and customer satisfaction level on customer loyalty. The testing uses the value of significance (sig) from the ANOVA output table. To see whether there is a linear relationship between service process, support facilities complaint handling, service quality and customer satisfaction level with customer loyalty we can perform the following analysis steps:

First: Determine the hypothesis to be tested
H0: There is no linear relationship between service process, support facilities complaint handling, service quality and customer satisfaction level with customer loyalty
H1: There is linear relationship between service process, support facilities complaint handling, service quality and customer satisfaction level with customer loyalty

Second: Calculate the value of significance (sig) or pvalue The sig value shown in column Sig is 0.000

Third: use the following criteria
If the sig value < 0.05 reject H0 and accept H1
If the sig value > 0.05 accept H0 and reject H1

Fourth: take the decision of the hypothesis testing

Since the sig value is 0.000 < 0.05; therefore reject H0 and accept H1. Thus there is linear relationship between service process, support facilities complaint handling, service quality and customer satisfaction level with customer loyalty. Accordingly service process, support facilities complaint handling, service quality and customer satisfaction level affect customer loyalty significantly.

The Partial Effect of Service Process, Support Facilities, Complaint Handling, Service Quality and Customer Satisfaction Variables on A Customer Loyalty Endogenous Variable

The magnitude of the effect of exogenous variables of service process, support facilities, complaint handling, service quality and customer satisfaction individually / partially on a customer loyalty endogenous variable can be seen from the Beta or Standardized Coefficients value. Whereas the testing of hypothesis uses t value. The values can be seen in the **Coefficient** table output. Therefore, in the following sections we will discuss the effect of exogenous variables of service process, support facilities, complaint handling , service quality and customer satisfaction on a customer loyalty endogenous variable partially.

Relationship between a Service Process Exogenous Variable and a Customer Loyalty Endogenous Variable

To see whether there is a linear relationship between a service process exogenous variable and a customer loyalty endogenous variable we can perform the following analysis steps:

First: Determine the hypothesis to be tested
H0: There is no linear relationship between a service process exogenous variable and a customer loyalty endogenous variable
H1: There is linear relationship between a service process exogenous variable and a customer loyalty endogenous variable

Second: Calculate the value of observation t (t_o)
The t_o value shown in column t in the above **Coefficients** table is -0.290

Third: Calculate the number of t table (t_α) as it has been discussed in the previous chapter

From these provisions it is obtained t_α as much as 1.960

Fourth: Define the decision-making criteria as it has been discussed in the previous chapter

Fifth: Make a decision of hypothesis testing results
The calculation result with IBM SPSS shows the t_o as much as -0.290. Since the t_o is negative, to make easier we can use the significance instead of t. The value of significance for the t_o in Column **Sig** is as much as 0.772. Since the sig from the calculation as much as 0.772 is bigger than 0.05 then there is no a significant influence of the relationship between the two variables.

Relationship between a Support Facilities Exogenous Variable and a Customer Loyalty Endogenous Variable

To see whether there is a linear relationship between a support facilities exogenous variable and a customer loyalty endogenous variable we can perform the following analysis steps:

First: Determine the hypothesis to be tested
H0: There is no linear relationship between a support facilities exogenous variable and a customer loyalty endogenous variable
H1: There is linear relationship between a support facilities exogenous variable and a customer loyalty endogenous variable

Second: Calculate the value of observation t (t_o)
The t_o value shown in column t in the above **Coefficients** table is 0.908

Third: Calculate the number of t table (t_α) as it has been discussed in the previous chapter

From these provisions it is obtained t_α as much as 1.960

Fourth: Define the decision-making criteria as it has been discussed in the previous chapter

Fifth: Make a decision of hypothesis testing results
The calculation result with IBM SPSS shows the t_o as much as 0.908 is smaller than t_α as much as 1.960; thus the decision is accept H0 and reject H1. This means that there is a no linear relationship between a support facilities exogenous variable and a customer loyalty endogenous variable.

Relationship between a Complaint Handling Exogenous Variable and a Customer Loyalty Endogenous Variable

To see whether there is a linear relationship between a complaint handling exogenous variable and a customer loyalty endogenous variable we can perform the following analysis steps:

First: Determine the hypothesis to be tested
H0: There is no linear relationship between a complaint handling exogenous variable and a customer loyalty endogenous variable
H1: There is linear relationship between a complaint handling exogenous variable and a customer loyalty endogenous variable

Second: Calculate the value of observation t (t_o)
The t_o value shown in column t in the above **Coefficients** table is 1.444

Third: Calculate the number of t table (t_α) as it has been discussed in the previous chapter

From these provisions it is obtained t_α as much as 1.960

Fourth: Define the decision-making criteria as it has been discussed in the previous chapter

Fifth: Make a decision of hypothesis testing results
The calculation result with IBM SPSS shows the t_o as much as 1.444 is smaller than t_α as much as 1.960; thus the decision is accept H0 and reject H1. This means that there is a no linear relationship between a support facilities exogenous variable and a customer loyalty endogenous variable.

Relationship between a Service Quality Exogenous Variable and a Customer Loyalty Endogenous Variable

To see whether there is a linear relationship between a service quality exogenous variable and a customer loyalty endogenous variable we can perform the following analysis steps:

First: Determine the hypothesis to be tested
H0: There is no linear relationship between a service quality exogenous variable and a customer loyalty endogenous variable
H1: There is linear relationship between a service quality exogenous variable and a customer loyalty endogenous variable

Second: Calculate the value of observation t (t_o)
The t_o value shown in column t in the above **Coefficients** table is 0.377

Third: Calculate the number of t table (t_α) as it has been discussed in the previous chapter

From these provisions it is obtained t_α as much as 1.960

Fourth: Define the decision-making criteria as it has been discussed in the previous chapter

Fifth: Make a decision of hypothesis testing results
The calculation result with IBM SPSS shows the t_o as much as 0.377 is smaller than t_α as much as 1.960; thus the decision is accept H0 and reject H1. This means that there is a no linear relationship between a support service quality variable and a customer loyalty endogenous variable.

PATH ANALYSIS: DATA ANALYSIS APPLICATION

Relationship between a Customer Satisfaction Exogenous Variable and a Customer Loyalty Endogenous Variable

To see whether there is a linear relationship between a customer satisfaction exogenous variable and a customer loyalty endogenous variable we can perform the following analysis steps:

First: Determine the hypothesis to be tested
H0: There is no linear relationship between a customer satisfaction exogenous variable and a customer loyalty endogenous variable
H1: There is linear relationship between a customer satisfaction exogenous variable and a customer loyalty endogenous variable

Second: Calculate the value of observation t (t_o)
The t_o value shown in column t in the above **Coefficients** table is 3.236

Third: Calculate the number of t table (t_α) as it has been discussed in the previous chapter

From these provisions it is obtained t_α as much as 1.960

Fourth: Define the decision-making criteria as it has been discussed in the previous chapter

Fifth: Make a decision of hypothesis testing results
The calculation result with IBM SPSS shows the t_o as much as 3.236 is bigger than t_α as much as 1.960; thus the decision is reject H0 and accept H1. This means that there is a linear relationship between a customer satisfaction exogenous variable and a customer loyalty endogenous variable. Because there is a linear relationship between the two variables; then the customer satisfaction affects the customer loyalty significantly. The amount of effect can be known from coefficient value of Beta (in column **Standardized Coefficient Beta**) as much as 0.344. The effect of this magnitude is significant because the value of significance in the **Sig column** is 0.002 which is smaller than 0.05.

Calculating Direct Effect (DE)

- The effect of complaint-handling on service quality X1 to Y = 0.503
- The effect of service facility support on service quality X2 to Y = 0.118
- The effect of service process on service quality X3 to Y = -0.015
- The effect of variable complaint handling to customer satisfaction X1 to Y2 = 0.460
- The effect of service facility support variable to customer satisfaction
X2 to Y2 = -0.114
- The effect of service process on customer satisfaction X3 to Y2 = 0.161
- The effect of complaints handling to customer loyalty X1 to Y3 = 0.281
- The effect of service support facility to customer loyalty X2 to Y3 = 0.240
- The effect of service process on customer loyalty X3 to Y3 = -0.053
- The effect of service quality to customer satisfaction Y1 to Y2 = 0.388
- The effect of customer satisfaction on customer loyalty Y2 to Y3 = 0.344

Calculating Indirect Effect (IE)

- The effect of complaint handling to customer satisfaction through service quality
X to Y2 through Y = (0.503 x 0.388) = 0.195
- The effect of service support facility on customer satisfaction through service quality
X2 to Y2 through Y = (0.118 x 0.388) = 0.045

- The effect of service process variables on customer satisfaction through service quality
  X3 to Y2 through Y = (-0.015 x 0.388) = -0.0058
- The effect of complaint handling to customer loyalty through service quality
  X to Y3 through Y = (0,503 x 0,034) = 0,017
- The effect of service support facility on customer loyalty through service quality
  X2 to Y3 through Y = (0.118 x 0.034) = 0.0040
- The effect of service process on customer loyalty through service quality
  X3 to Y3 through Y = (-0,015 x 0,034) = -0,0005
- The effect of complaint handling on customer loyalty through customer satisfaction
  X to Y3 through Y2 = (0.460 x 0.344) = 0.158
- The effect of service support facility on customer loyalty through customer satisfaction
  X to Y3 through Y2 = (0.118 x 0.344) = 0.040
- The effect of service process on customer loyalty through customer satisfaction
  X to Y3 through Y2 = (0.161 x 0.344) = 0.055
- The effect of complaint handling on customer loyalty through service quality and customer satisfaction X to Y3 through Y1 and Y2 = (0.503 x 0.388 x 0.344) = 0.067
- The effect of service support facility on customer loyalty through service quality and customer satisfaction X2 to Y3 through Y1 and Y2 = (0.118 x 0.388 x 0.344) = 0.015
- The effect of service process on customer loyalty through service quality and customer satisfaction X to Y3 through Y1 and Y2 = (-0.015 x 0.388 x 0.344) = -0.002

Total Effect (TE)

- The effect of complaint handling to customer satisfaction through service quality
  X1 to Y2 through Y = (0.503 + 0.388) = 0.891
- The effect of service support facility on customer satisfaction through service quality
  X2 to Y2 through Y = (0.118 + 0.388) = 0.506
- The effect of service process on customer satisfaction through service quality
  X3 to Y2 through Y = (-0.015 + 0.388) = 0.373
- The effect of complaint handling on customer loyalty through service quality
  X1 to Y3 through Y = (0.503 + 0.034) = 0.537
- The effect of service support facility on customer loyalty through service quality
  X2 to Y3 through Y = (0.118 + 0.034) = 0.152
- The effect of service process on customer loyalty through service quality
  X3 to Y3 through Y = (-0.015 + 0.034) = 0.019
- The effect of complaint handling to customer loyalty through customer satisfaction
  X to Y3 through Y2 = (0.460 + 0.344) = 0.804
- The effect of service support facility on customer loyalty through customer satisfaction
  X to Y3 through Y2 = (0.118 + 0.344) = 0.462
- The effect of service process on customer loyalty through customer satisfaction
  X to Y3 through Y2 = (0.161 + 0.344) = 0.05
- The effect of complaint handling on customer loyalty through service quality and customer satisfaction.
  X to Y3 through Y1 and Y2 = (0.503 + 0.388 + 0.344) = 1.235
- The effect of service support facility on customer loyalty through service quality and customer satisfaction
  X2 to Y3 through Y1 and Y2 = (0.118 + 0.388 + 0.344) = 0.85

PATH ANALYSIS: DATA ANALYSIS APPLICATION

- The effect of service process on customer loyalty through service quality and customer satisfaction
X to Y3 through Y1 and Y2 = (-0.015 + 0.388 + 0.344) = 0.717

The Path Diagram of the Complex Model

The Path diagram of the complex model is as follows:

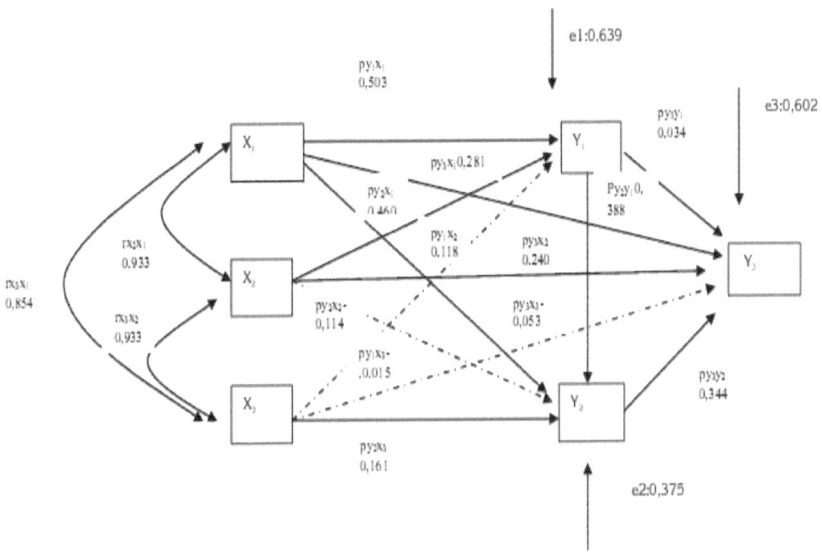

Figure 8.4 A Complex Path Diagram of Parameter Values on Complaint Handling, Service Support, Service Process, Service Quality, Customer Satisfaction and Customer Loyalty Variables' Relationship

Structural equations for the complex model above are:
sub structure 1: Y1 = 0.503 X1 + 0.118 X2 - 0.015 X3 + e 1
sub structure 2: Y2 = 0.460 X1 - 0.114 X2 + 0.161 X3 + 0.388Y1 + e 2
sub structure3: Y3 = 0.281 X1 + 0.240 X2 - 0.053 X3 + 0.344 Y2 + 0.034 Y1 + e 3

CHAPTER 9

PATH ANALYSIS USING PANEL DATA

9.1 Definition of Panel Data

Panel data is combined data originating from time series data and cross – section one. Time series data is data in which every observation is identified using time or date. While cross – section data is data in which every observation is identified by unique ID, such as name of companies and name of countries or cities. Panel data is also called as "pooled data" possessing spatial and time dimensions. Below it will be shown examples of time series data and cross – section data as well as panel data.

Time Series Data

The following is an example of time series data of sales.

Year	Sales
2000	1000000
2001	1020000
2002	1023000
2003	1203000
2004	1320000
2005	1420000

PATH ANALYSIS: DATA ANALYSIS APPLICATION

Cross Section Data

The following is an example of cross section data of customers from an ABC company in 2017

Number	Names of Customers
1	Ann
2	Emily
3	John
4	Cloe
5	Brian

Panel Data

The following is an example of panel data on number of employees and customers for companies of A,B,C and D from 2015 – 2017.

Companies	Years	Number of Employees	Number of Customers
A	2015	100	1000
	2016	100	1500
	2017	150	2000
B	2015	103	1200
	2016	104	1250
	2017	110	3000
C	2015	110	1400
	2016	115	2000
	2017	120	3500
D	2015	98	1100
	2016	100	1500
	2017	125	3000

In this example, we will use two independent variables that function as exogenous variables and one variable that serves as an intervening variable and one dependent variable that serves as an endogenous variable. As the exogenous variable are X1 and X2, while the intervening variable is X3 and as the endogenous variable is Y. The relationship between these variables can be described in the path diagram model as follows:

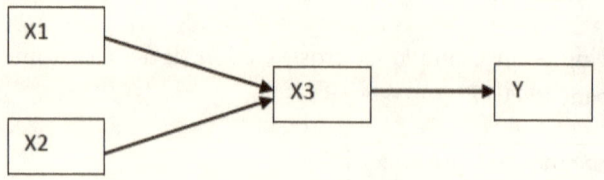

Figure 9.1 Path Diagram Model of X1, X2, X3 and Y Variables Based on Theory

Hypotheses and Problems
The first hypothesis for this case is:
H0: X1 and X2 exogenous variables have no effect on X3 significantly either simultaneously or partially
H1: X1 and X2 exogenous variables have effect on X3 significantly either simultaneously or partially

The second hypothesis for this case is:
H0: X1 and X2 exogenous variables have no effect on Y significantly through X3
H1: X1 and X2 exogenous variables have effect on Y significantly through X3

The problems for this case are as follows:
a. How much do the influence of X1 and X2 variables to X3 partially?
b. How much do the influence of X1 and X2 variables on X3 simultaneously?
c. How much do the influence of X1 variable to Y through X3?
d. How much do the influence of X2 variable to Y through X3?

PATH ANALYSIS: DATA ANALYSIS APPLICATION

Research Data

The research data of this case is as follows:

Year	ID	Y	X1	X2	X3
2005	1	90	98	97	95
2006	1	100	105	102	100
2007	1	80	85	83	81
2008	1	95	90	94	92
2009	1	98	99	97	95
2010	1	89	90	88	86
2011	1	86	90	85	83
2005	2	87	91	86	84
2006	2	105	100	104	102
2007	2	135	141	133	130
2008	2	150	151	149	147
2009	2	75	78	74	72
2010	2	78	81	77	75
2011	2	69	71	68	62
2005	3	68	70	67	65
2006	3	66	68	65	63
2007	3	64	66	64	62
2008	3	60	62	61	60
2009	3	58	60	60	58
2010	3	56	59	57	55
2011	3	52	54	53	51
2005	4	50	52	51	50
2006	4	48	50	49	47
2007	4	44	46	45	43
2008	4	41	43	42	41
2009	4	38	40	40	39
2010	4	36	40	38	46
2011	4	34	36	36	35
2005	5	31	33	33	32
2006	5	28	30	30	29
2007	5	25	28	26	25
2008	5	21	25	23	21
2009	5	18	20	20	19
2010	5	14	18	19	18
2011	5	11	16	15	14

Input the data above in Excell and give name as pa_data. xlsx

To solve the case the stages are as follows:

First: Creating a path diagram model based on the relationship between the variables we are examining based on theory.

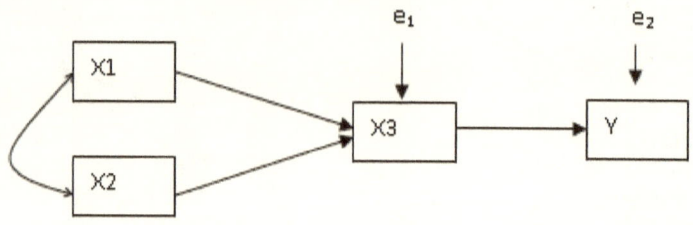

Figure 9.2 Path Diagram Model of X1, X2, X3 and Y Variables

Second: Creating a path diagram based on the relationship between the variables we are examining

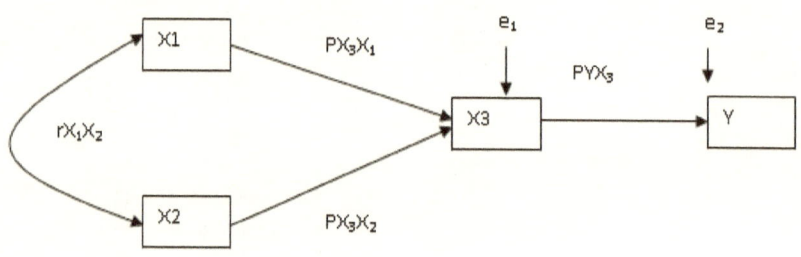

Figure 9.3 Path Diagram of X1, X2, X3 and Y Variables

Where:
X1 as an exogenous variable
X2 as an exogenous variable
X3 as an endogenous intervening variable
Y1 as an endogenous variable

PATH ANALYSIS: DATA ANALYSIS APPLICATION

Third: Create a structural equation
The path diagram above has two structural equations. The structural equation is as follows:

- X3 = PX31X1 + PX3X + e1 (As a sub-structural equation 1)
- Y = PYX + e (As a sub-structural equation 2)

Fourth: Analyze the data using Stata

To analyze the data we need to conduct the following steps: First step: Convert the data into Stata format using Stat Transfer program (*You must have **Stat Transfer** program first*). The steps to convert the data are as shown in the box below. Save the Stata file with the name pa_data.dta

---

- Activate the **Stat Transfer** program
- Select Excel in the Input File Type option. In the File Specification option: locate the file pa_data.xlsx to be converted by clicking the Browse command
- In the Output File Type option select Stata
- In the File Specification option at the bottom find the location where we will save the converted file with the name as pa_data.dta
- Click OK

  The data is ready to be analysed using Stata.

Second step: Analyze data using Stata
Calculating Sub Structure 1: Effect of X1 and X2 exogenous variables to endogenous X3 variable. The steps to calculate:

- Activate **Stata**
- **File> Open> Open** pa_data.dta**> Open**
- Select **Statistics> Linear Model and Related> Data Panel> Linear Regression**
- Enter the X3 Variable into the **Dependent Variable** column as well as the X1 and X2 variables into **Independent Variables**
- In the **Type menu** select by ticking (v) on **GLS Random Effect**
- **Select Panel Settings**
- **Panel ID Variable**: select id_
- Activate the option **Time Variable** by putting a checklist (v) then select years
- Check **Yearly** on the option **Time Unit** and the display format for the **Time Variable**
- click **OK**

The result of calculation is as follows

```
. xtreg x3 x1 x2, re

Random-effects GLS regression              Number of obs      =         33
Group variable: id_                        Number of groups   =          5

R-sq:                                      Obs per group:
    within  = 0.9868                                    min =          6
    between = 0.9996                                    avg =        6.6
    overall = 0.9967                                    max =          7

                                           Wald chi2(2)       =    7170.90
corr(u_i, X)  = 0 (assumed)                Prob > chi2        =     0.0000
```

x3	Coef.	Std. Err.	z	P>|z|	[95% Conf. Interval]	
x1	-.0210745	.1520413	-0.14	0.890	-.3190701	.2769211
x2	1.004534	.1557233	6.45	0.000	.6993218	1.309746
_cons	-.4477992	.8424055	-0.53	0.595	-2.098884	1.203285
sigma_u	.48063739					
sigma_e	1.8687983					
rho	.06204305	(fraction of variance due to u_i)				

PATH ANALYSIS: DATA ANALYSIS APPLICATION

The important values from the above output:

- R square (R^2) on the **R-sq** consists of the value of R square within of 0.9868; the value of the between R square value of 0.9996 and overall R square as much as 0.9967.
- The value of the Chi Square ($\chi2$) on the part of **Wald chi2** as much as 7170.90 with probability (**Prob > chi2**) of 0.0000.
- The value of regression coefficient for X 1 variable in column **Coef** is as much as - 0.0210745 with probability values in column **P > | z |** of 0.890
- The value of regression coefficient for X 2 variable in column **Coef** is as much as 1.004534 with probability values in column **P > | z |** of 0.000.

Note that the output of regression coefficient on Stata is the value of the unstandardized regression coefficients (b). Meanwhile for the procedure of path analysis uses standardized regression coefficients (Beta). Accordingly, before that value is used in the path analysis we need to convert that value into standardized regression coefficients. To calculate the value of the standardized regression coefficients can be done as follows: standardized regression coefficients is multiplied by the result of standard deviation of the independent variable which is divided by the standard deviation of the dependent variable.

Change the value of unstandardized regression coefficients (b) into standardized regression coefficients (β)

The steps to change the value of unstandardized regression coefficients (b) into standardized regression coefficients (β) is as follows:

First: calculate the value of the standard deviation of all the variables examined in Stata with the following way

- Select **Statistics > Summaries, Tables and Tests > Summaries and Descriptives Statistics > Summary**
- On the option of **Variables**, select Y, x 1, x 2, and X 3
- On the choice of **Options**, select the **Standard Display**
- Click OK

The result will be like this

```
summarize y x1 x2 x3
```

Variable	Obs	Mean	Std. Dev.	Min	Max
y	33	63.84848	33.59048	11	150
x1	33	66.30303	33.36024	16	151
x2	33	64.78788	32.53148	15	149
x3	33	63.24242	31.95997	14	147

The values of the standard deviation of the variables we studying are as follows:

- the value of the standard deviation of the Y dependent variable) is 33.59048
- the value of the standard deviation of the X1 independent variable is 33.36024
- the value of the standard deviation of the X2 independent variable is 32.53148
- the value of the standard deviation of the X 3 intervening variable is 31.95997

Second: Calculate with following formula to obtain the values of standardized regression coefficients

$$\text{standardized regression coefficient} = \text{unstandardized regression coefficient} \times \frac{\text{standard deviation of the independent variable}}{\text{standard deviation of the dependent variable}}$$

Standardized regression coefficients for sub structure 1 is: standardized regression coefficients of X1 and X2. Standardized regression coefficients for sub structure 2 is: standardized regression coefficients of X1, X2 and X3.

Standardized regression coefficient of X1 can be obtain as follows:

PATH ANALYSIS: DATA ANALYSIS APPLICATION

$$\text{Standardized regression coefficient of X1} = -0,0210745 \times \frac{33,36024}{31,95997}$$

From the calculation, we get the value of X1 standardized regression coefficient as much as -0.02199784

Standardized regression coefficient of X2 can be obtain as follows:

$$\text{standardized regression coefficient of X2} = 1,004534 \times \frac{32,53148}{31,95997}$$

From the calculation, we get the value of X2 standardized regression coefficient as much as 1,0223

Standardized regression coefficient of X3 can be obtain as follows:

$$\text{standardized regression coefficient of X3} = 1,040398 \times \frac{31,95997}{33,59048}$$

From the calculation, we get the value of X3 standardized regression coefficient as much as 0,989896.

Third: use the value of the standardized regression coefficient as path coefficients as follows:

- Standardized regression coefficient of X1 to X3 is -002199784
- Standardized regression coefficient of X2 to X3 is 1,023

Fourth: Interpret the value of the outputs

- The value of R square (R^2) on the column from the output (**R-sq**) on top of 0.9967. This value has the meaning that magnitude of variation of X 3 endogenous Variable which can be described using X 1 and X 2 exogenous Variables is as much as 0,9967; while the remainder of 0.0033 is influenced by factors other than this research.
- A path coefficient of an X 1 exogenous variable to the X 3 endogenous variable is:-0.02199784 with probability values in column **P > | z |** of 0.890 meaning that the X 1 has no significant effect on X 3 because the value of the probability of 0.890 > 0.05.
- A path coefficient of an X 2 exogenous variable to the X 3 endogenous variable is: 1.023 with probability values in column **P > | z |** of 0.000 meaning that the X 2 has significant effect on X 3 because the value of the probability of 0.000 < 0.05.

Fifth: calculating correlation of exogenous variables of X 1 and X 3 as follows:
- **Select Statistics > Summaries, Tables and Tests > Summaries and Descriptives Statistics > Correlations and Covariances**
- On the choice of **Variables**: select x 1 and X 2 > Click **OK**

The result will be like this

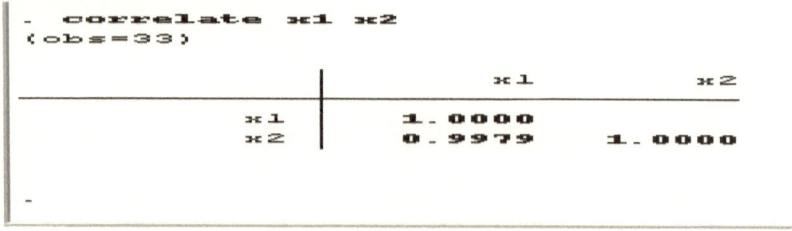

The output shows that the magnitude of the correlation between X 1 and X 2 is 0.9979 which has the meaning of that the correlation between the two variables is very strong.

PATH ANALYSIS: DATA ANALYSIS APPLICATION

Calculate the Sub Structure II: X 3 to Y

Use the following steps
- Activate Stata
- File Open Open pa_data. dta > Open
- Select the Statistics of Linear Models and Related > Panel Data > Linear Regression
- **Select Statistics> Linear Model and Related> Data Panel> Linear Regression**
- Enter the Y Endogenous Variable into the **Dependent Variable** column and X3 variable into **Independent Variables**
- In the **Type menu** select by ticking (v) on **GLS Random Effect**
- **Select Panel Settings**
- **Panel ID Variable:** select id _
- Activate the option **Time Variable** by putting a checklist (v) then select years
- Check **Yearly** on the option **Time Unit** and the display format for the **Time Variable**
  click **OK**

```
. xtreg y x3, re

Random-effects GLS regression                   Number of obs      =       33
Group variable: id_                             Number of groups   =        5

R-sq:                                           Obs per group:
     within  = 0.9742                                        min =        6
     between = 0.9989                                        avg =      6.6
     overall = 0.9933                                        max =        7

                                                Wald chi2(1)       =  3372.36
corr(u_i, X)   = 0 (assumed)                    Prob > chi2        =   0.0000

------------------------------------------------------------------------------
           y |      Coef.   Std. Err.      z    P>|z|     [95% Conf. Interval]
-------------+----------------------------------------------------------------
          x3 |   1.040398    .0179157    58.07   0.000     1.005284    1.075512
       _cons |  -1.947142    1.274783    -1.53   0.127    -4.445671     .5513869
-------------+----------------------------------------------------------------
     sigma_u |  .80330139
     sigma_e |  2.5736589
         rho |  .08877314   (fraction of variance due to u_i)
------------------------------------------------------------------------------
```

The results of the calculation is as follows:
- The value of R square (R^2) on the column from the output (**R-sq**) on top of 0.9933. This value has the meaning that magnitude of variation of a Y endogenous Variable which can be described using an X 3 exogenous Variable is as much as 0,9933; while the remainder of 0.0067 is influenced by factors other than this research
- A path coefficient of an X 3 exogenous variable to t a Y endogenous variable is 0,989896 with probability values in column **P > | z |** of 0.000 meaning that the X 3 has significant effect on Y because the value of the probability of 0.000 < 0.05.

Draw a Path Diagram of values already found from the calculation

The path diagram of the model we have studied is as follows:

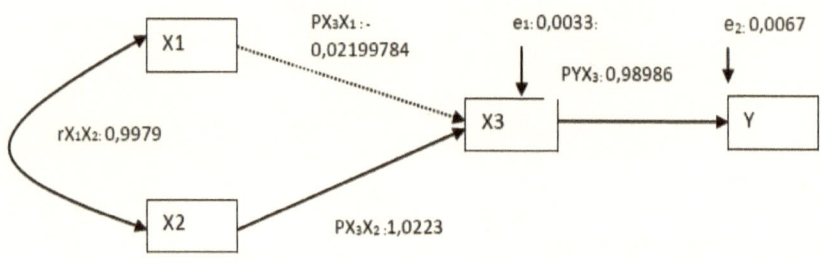

Make Conclusion of the research we have done

The conclusion of the research is as follows:

- The effect of X_1 on X_3 is -0.02199784 and the effect is not significant
- The effect of X_2 on X_3 is 1.023 and the effect is significant
- The effect of X_3 on Y is 0.98986 and the effect is significant
- The correlation between X_1 and X_2 is 0,9979
- The effect of X_1 on Y through X_3 is -0,02199784 x 0,98986 = -0,02177
- The effect of X_2 on Y through X_3 is 1,0223 x 0,98986 = 1,011934
- The effect of X_1 and X_3 on Y is -0,02199784 + 0,98986 = 0,967862
- The effect of X_2 and X_3 on Y is 1,0223 + 0,98986 = 2,01216
- The effect of X_1 and X_2 on X_3 simultaneously is 0,9967
- The effect of X_1, X_2 and X_3 on Y simultaneously is 0,9933

CHAPTER 10

AN APPLICATION IN THESIS RESEARCH

10.1 Research Problems

In this example, the writer uses a research that has been conducted by the graduate student whose topic is "Effect of Usefulness and Easiness Perception of E-filing and its Impact on Taxpayers' Obedience in Central Jakarta". As the exogenous variable is perception of use and the endogenous one is taxpayers' obedience. While the perception of easiness variable is treated as an intervening variable. The model of the relationship among the variables studied is as follows.

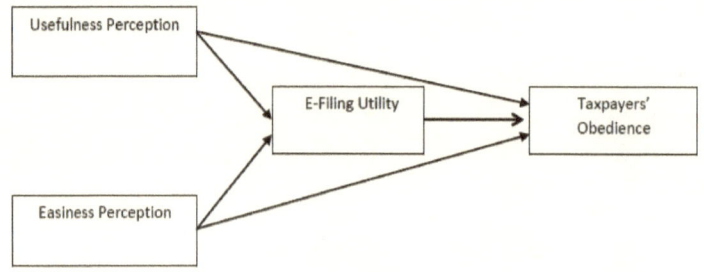

Figure 10.1 Relationship among the Variables of Usefulness Perception, Easiness Perception, E-filing Utility and Taxpayers Based on Theory

The problem formulation of this research is as follows:
1. Does usefulness perception affect e-filing utility?
2. Does easiness perception affect e-filing utility?
3. Does usefulness perception affect the taxpayers' obedience?
4. Does easiness perception affect the taxpayers' obedience?
5. Does e-filing utility affect the taxpayers' obedience?
6. Does usefulness perception affect the taxpayers' obedience through e-filing utility?
7. Does easiness perception affect the taxpayers' obedience through e-filing utility?

10.2 Methodology

This research uses a quantitative approach, descriptive design and survey method. The population of the research is individual taxpayers in West Jakarta. The sample used in the study as much as 390 respondents is drawn using a probability technique. Path analysis procedure is used to analyze the primary data that have been collected using questionnaires. The following is the path analysis model of relationship among the variables of usefulness perception, easiness perception, e-filing utility and taxpayers

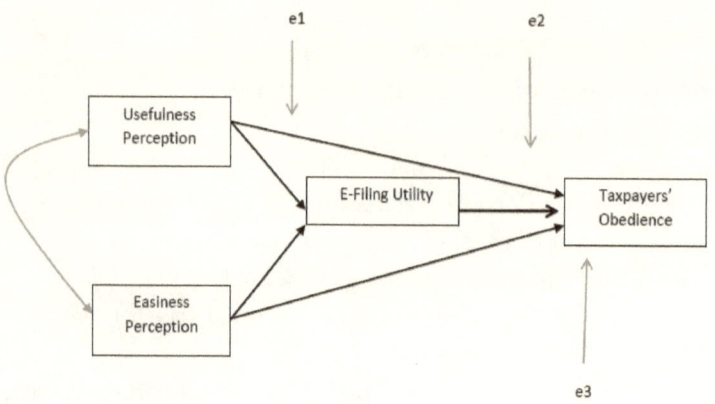

Figure 10.2 A Path Diagram Model of Relationship among the Variables of Usefulness Perception, Easiness Perception, E-filing Utility and Taxpayers

The above path analysis model have a path diagram as follows:

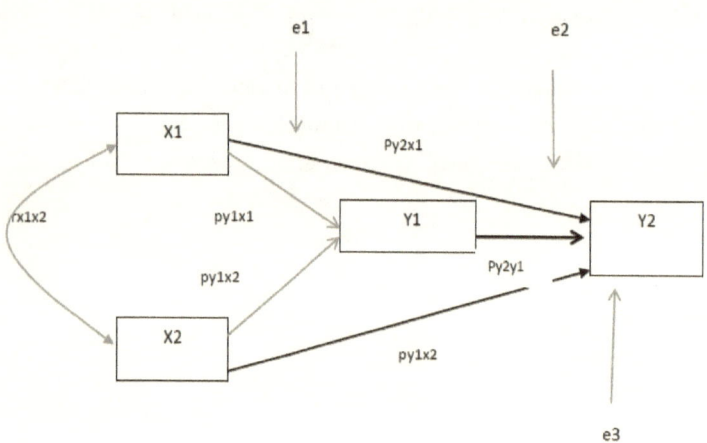

Figure 10.3 A Path Diagram of Relationship among the Variables of Usefulness Perception, Easiness Perception, E-filing Utility and Taxpayers

PATH ANALYSIS: DATA ANALYSIS APPLICATION

Where:

- X1: Usefulness Perception
- X2: Easiness Perception
- Y1: E-Filing Utility
- Y2: Taxpayers' Obedience

The path diagram above has three structural equations. The structural equation can be seen as below:
- Y1 = PY1X1 + PY1 X2 + e1 (As a sub-structure equation 1)
- Y2 = PY2 Y1 + e2 (As a sub-structure equation 2)
- Y3 = PY2X1 + PY2 X2 + e3 (As a sub-structure equation 3)

10.3 Research Result

Research results based on the field data are as follows:

First Sub – Structure

The next discussion is the result of the first structure calculation. In this part, we will discuss both the combined effect and partial effect of exogenous usefulness perception and easiness perception variables on a e-filing utility variable.

a. Dependent Variable: y1

b. All requested variables entered.

Model Summary[b]

Model	R	R Square	Adjusted R Square	Std. Error of the Estimate	Durbin-Watson
1	.613[a]	.375	.372	3.197047	2.087

The Combined Effect of Exogenous Usefulness Perception and Easiness Perception Variables on a E-Filing Utility Variable

The combined effect of exogenous usefulness perception and easiness perception variables on a e-filing utility variable can be seen in the output in the **Model Summary** table, in the R square value. The value of R square (R^2) in the table above is 0.375. This number shows the amount of variation of the e-filing utility variable that can be explained by exogenous variables of usefulness perception and easiness perception. It can also be said that the R square (R^2) of 0.375 shows the amount of the effect of exogenous variables of usefulness perception and easiness perception on the endogenous e-filing utility variable. The R square can also be stated in the form Coefficient of Determination. In this case the amount of the Coefficient of Determination is as much as 37.5%. While the remainder as much as 62.5% is due to the extraneous variables.

Simultaneous Hypothesis Testing

This hypothesis testing is used to see the simultaneous effect of usefulness perception and easiness perception variables on a e-filing utility variable. The testing uses the value of significance (sig) from the ANOVA output table.

ANOVA[a]

Model		Sum of Squares	df	Mean Square	F	Sig.
1	Regression	2375.249	2	1187.624	116.193	.000[b]
	Residual	3955.570	387	10.221		
	Total	6330.819	389			

Relationship between Usefulness Perception and Easiness Perception with E-Filing Utility

To see whether there is a linear relationship between usefulness perception and easiness perception with e-filing utility we can perform the following analysis steps:

PATH ANALYSIS: DATA ANALYSIS APPLICATION

First: Determine the hypothesis to be tested
H0: There is no linear relationship between usefulness perception and easiness perception with e-filing utility
H1: There is linear relationship between usefulness perception and easiness perception with e-filing utility

Second: Calculate the value of significance (sig) or pvalue
The sig value shown in column Sig is 0.000

Third: use the following criteria
If the sig value < 0.05 reject H0 and accept H1
If the sig value > 0.05 accept H0 and reject H1

Fourth: take the decision of the hypothesis testing

Since the sig value is 0.000 < 0.05; therefore reject H0 and accept H1. Thus there is linear relationship between usefulness perception and easiness perception with e-filing utility. Accordingly usefulness perception and easiness perception affect e-filing utility significantly.

The Partial Effect of Usefulness Perception and Easiness Perception Variables on a E-Filing Utility Variable

In the following part, we will discuss partial effect of exogenous usefulness perception and easiness perception variables on a e-filing utility variable one by one.

Coefficients^a

Model		Unstandardized Coefficients		Standardized Coefficients	t	Sig.	Collinearity Statistics
		B	Std. Error	Beta			Tolerance
1	(Constant)	7.460	1.211		6.159	.000	
	x1	.140	.047	.156	2.956	.003	.578
	x2	.545	.058	.499	9.444	.000	.578

Relationship between Usefulness Perception and E-Filing Utility

To see whether there is a linear relationship between usefulness perception and e-filing utility we can perform the following analysis steps:
First: Determine the hypothesis to be tested
H0: There is no linear relationship between usefulness perception and e-filing utility
H1: There is linear relationship usefulness perception and e-filing utility

Second: Calculate the value of observation t (t_o)
The t_o value shown in column t in the above **Coefficients** table is 2.956
Third: Calculate the number of t table (t_α)
The t_α is as much as 1.960

Fourth: Define the decision-making criteria as it has been discussed above.

Fifth: Make a decision of hypothesis testing results

The calculation result with IBM SPSS shows the t_o as much as 2.956 is bigger than t_α as much as 1.960; thus the decision is reject H0 and accept H1. This means that there is a linear relationship between usefulness perception and e-filing utility. Because there is a linear relationship between the two variables; then the usefulness perception affects e-filing utility significantly. The amount of effect can be known from coefficient value of Beta (in column **Standardized Coefficient Beta**) as much as 0.156. The effect of this magnitude is significant because the value of significance in the **Sig column** is 0.003 which is smaller than 0.05.

Relationship between Easiness Perception and E-Filing Utility

To see whether there is a linear relationship between easiness perception and e-filing utility we can perform the following analysis steps:
First: Determine the hypothesis to be tested
H0: There is no linear relationship between easiness perception and e-filing utility
H1: There is linear relationship easiness perception and e-filing utility

PATH ANALYSIS: DATA ANALYSIS APPLICATION

Second: Calculate the value of observation t (t_o)
The t_o value shown in column t in the above **Coefficients** table is 9.444

Third: Calculate the number of t table (t_α)
The t_α is as much as 1.960

Fourth: Define the decision-making criteria as it has been discussed above.

Fifth: Make a decision of hypothesis testing results

The calculation result with IBM SPSS shows the t_o as much as 9.444 is bigger than t_α as much as 1.960; thus the decision is reject H0 and accept H1. This means that there is a linear relationship between easiness perception and e-filing utility. Because there is a linear relationship between the two variables; then the easiness perception affects e-filing utility significantly. The amount of effect can be known from coefficient value of Beta (in column **Standardized Coefficient Beta**) as much as 0.499. The effect of this magnitude is significant because the value of significance in the **Sig column** is 0.000 which is smaller than 0.05.

Second Sub – Structure

In this second structure we will discuss the effect of e-filing utility on taxpayer's obedience.

Model	R	R Square	Adjusted R Square	Std. Error of the Estimate	Durbin-Watson
1	.716ª	.512	.511	1.785717	2.210

The effect of e-filing utility on taxpayer's obedience can be seen in the output in the **Model Summary** table, in the R square value. The value of R square (R^2) in the table above is 0.512. This number shows the amount of variation of taxpayer's obedience that can be explained by e-filing utility. It can also be said that the R square (R^2) of 0.512 shows the amount of the effect of e-filing utility on taxpayer's obedience. The R square can also be stated in the form Coefficient of Determination. In this case the amount of the Coefficient of Determination is as much as 51.2%. While the remainder as much as 48.8% is due to the extraneous variables.

ANOVA^a

Model		Sum of Squares	df	Mean Square	F	Sig.
1	Regression	1300.278	1	1300.278	407.766	.000^b
	Residual	1237.248	388	3.189		
	Total	2537.526	389			

Relationship between E-Filing Utility and Taxpayer's Obedience

To see whether there is a linear relationship between e-filing utility and taxpayer's obedience we can perform the following analysis steps:

First: Determine the hypothesis to be tested

H0: There is no linear relationship between e-filing utility and taxpayer's obedience

H1: There is linear relationship between e-filing utility and taxpayer's obedience

Second: Calculate the value of significance (sig) or pvalue
The sig value shown in column Sig is 0.000
Third: use the following criteria
If the sig value < 0.05 reject H0 and accept H1
If the sig value > 0.05 accept H0 and reject H1

Fourth: take the decision of the hypothesis testing

Since the sig value is 0.000 < 0.05; therefore reject H0 and accept H1. Thus there is linear relationship between e-filing utility and taxpayer's obedience. Accordingly e-filing utility affects taxpayer's obedience significantly.

PATH ANALYSIS: DATA ANALYSIS APPLICATION

Coefficients

Model		Unstandardized Coefficients		Standardized Coefficients	t	Sig.	Collinearity Statistics
		B	Std. Error	Beta			Tolerance
1	(Constant)	7.629	.585		13.040	.000	
	Y1	.453	.022	.716	20.193	.000	1.000

To see whether there is a linear relationship between easiness perception and e-filing utility we can perform the following analysis steps:

First: Determine the hypothesis to be tested

H0: There is no linear relationship between e-filing utility and taxpayer's obedience

H1: There is linear relationship between e-filing utility and taxpayer's obedience

Second: Calculate the value of observation t (t_o)

The t_o value shown in column t in the above **Coefficients** table is 20.193

Third: Calculate the number of t table (t_α)

The t_α is as much as 1.960

Fourth: Define the decision-making criteria as it has been discussed above.

Fifth: Make a decision of hypothesis testing results

The calculation result with IBM SPSS shows the t_o as much as 20.193 is bigger than t_α as much as 1.960; thus the decision is reject H0 and accept H1. This means that there is a linear relationship between e-filing utility and taxpayer's obedience. Because there is a linear relationship between the two variables; then the e-filing utility affects taxpayer's obedience significantly. The amount of effect can be known from coefficient value of Beta (in column **Standardized Coefficient Beta**) as much as 0.716. The effect of this magnitude is significant because the value of significance in the **Sig column** is 0.000 which is smaller than 0.05. The value as much as 0.716

means the increase of the taxpayer's obedience when the e-filing utility shows one unit increase.

Third Sub – Structure

The next discussion is the result of the first structure calculation. In this part, we will discuss both the combined effect and partial effect of exogenous usefulness perception and easiness perception variables on a taxpayer's obedience variable.

Model Summary[b]

Model	R	R Square	Adjusted R Square	Std. Error of the Estimate	Durbin-Watson
1	.697[a]	.486	.483	1.836469	2.219

The Combined Effect of Exogenous Usefulness Perception and Easiness Perception Variables on a Taxpayer's Obedience Variable

The combined effect of exogenous usefulness perception and easiness perception variables on a taxpayer's obedience variable can be seen in the output in the **Model Summary** table, in the R square value. The value of R square (R^2) in the table above is 0.486. This number shows the amount of variation of the taxpayer's obedience variable that can be explained by exogenous variables of usefulness perception and easiness perception. It can also be said that the R square (R^2) of 0.486 shows the amount of the effect of exogenous variables of usefulness perception and easiness perception on the endogenous taxpayer's obedience variable. The R square can also be stated in the form Coefficient of Determination. In this case the amount of the Coefficient of Determination is as much as 48.6%. While the remainder as much as 51.4% is due to the extraneous variables.

Simultaneous Hypothesis Testing

This hypothesis testing is used to see the simultaneous effect of usefulness perception and easiness perception variables on a taxpayer's obedience variable. The testing uses the value of significance (sig) from the

ANOVA output table.

ANOVAª

Model		Sum of Squares	Df	Mean Square	F	Sig.
1	Regression	1232.324	2	616.162	182.695	.000ᵇ
	Residual	1305.203	387	3.373		
	Total	2537.526	389			

Relationship between Usefulness Perception and Easiness Perception with Taxpayer's Obedience

To see whether there is a linear relationship between usefulness perception and easiness perception with taxpayer's obedience we can perform the following analysis steps:

First: Determine the hypothesis to be tested
H0: There is no linear relationship between usefulness perception and easiness perception with taxpayer's obedience
H1: There is linear relationship between usefulness perception and easiness perception with taxpayer's obedience

Second: Calculate the value of significance (sig) or pvalue
The sig value shown in column Sig is 0.000

Third: use the following criteria
If the sig value < 0.05 reject H0 and accept H1
If the sig value > 0.05 accept H0 and reject H1

Fourth: take the decision of the hypothesis testing

Since the sig value is 0.000 < 0.05; therefore reject H0 and accept H1. Thus there is linear relationship between usefulness perception and easiness perception with taxpayer's obedience. Accordingly usefulness perception and easiness perception affect taxpayer's obedience significantly.

The Partial Effect of Usefulness Perception and Easiness Perception Variables on a Taxpayer's Obedience Variable

In the following part, we will discuss partial effect of exogenous usefulness perception and easiness perception variables on a taxpayer's obedience variable one by one.

Coefficients^a

Model		Unstandardized Coefficients B	Unstandardized Coefficients Std. Error	Standardized Coefficients Beta	t	Sig.	Collinearity Statistics Tolerance
1	(Constant)	6.122	.696		8.800	.000	
	X1	.104	.027	.182	3.799	.000	.578
	X2	.390	.033	.565	11.768	.000	.578

Relationship between Usefulness Perception and Taxpayer's Obedience

To see whether there is a linear relationship between usefulness perception and taxpayer's obedience we can perform the following analysis steps:
First: Determine the hypothesis to be tested
H0: There is no linear relationship between usefulness perception and taxpayer's obedience
H1: There is linear relationship usefulness perception and taxpayer's obedience

Second: Calculate the value of observation t (t_o)
The t_o value shown in column t in the above **Coefficients** table is 3.799
Third: Calculate the number of t table (t_α)
The t_α is as much as 1.960

Fourth: Define the decision-making criteria as it has been discussed above.

Fifth: Make a decision of hypothesis testing results

The calculation result with IBM SPSS shows the t_o as much as 3.799 is bigger than t_α as much as 1.960; thus the decision is reject H0 and accept H1. This means that there is a linear relationship between usefulness perception and taxpayer's obedience. Because there is a linear relationship between the two variables; then the usefulness perception affects taxpayer's obedience significantly. The amount of effect can be known from coefficient value of Beta (in column **Standardized Coefficient Beta**) as much as 0.182. The effect of this magnitude is significant because the value of significance in the **Sig column** is 0.000 which is smaller than 0.05.

Relationship between Easiness Perception and Taxpayer's obedience

To see whether there is a linear relationship between easiness perception and taxpayer's obedience we can perform the following analysis steps:

First: Determine the hypothesis to be tested
H0: There is no linear relationship between easiness perception and taxpayer's obedience
H1: There is linear relationship easiness perception and taxpayer's obedience

Second: Calculate the value of observation t (t_o)
The t_o value shown in column t in the above **Coefficients** table is 11.768
Third: Calculate the number of t table (t_α)
The t_α is as much as 1.960

Fourth: Define the decision-making criteria as it has been discussed above.

Fifth: Make a decision of hypothesis testing results

The calculation result with IBM SPSS shows the t_o as much as 11.768 is bigger than t_α as much as 1.960; thus the decision is reject H0 and accept H1. This means that there is a linear relationship between easiness perception and taxpayer's obedience. Because there is a linear relationship between the two variables; then the easiness perception affects taxpayer's obedience significantly. The amount of effect can be known from coefficient value of Beta (in column **Standardized Coefficient Beta**) as much as 0.565. The effect of this magnitude is significant because the value of significance in the **Sig column** is 0.000 which is smaller than 0.05.

Conclusion

From the calculation above, the conclusion of the research is as follows:

1. Usefulness perception affects e-filing utility significantly as much as 0.156
2. Easiness perception affects e-filing utility significantly as much as 0.499
3. Usefulness perception affects the taxpayers' obedience significantly as much as 0.182
4. Easiness perception affects the taxpayers' obedience significantly as much as 0.565
5. E-filing utility affect the taxpayers' obedience as much as 0.716
6. Usefulness perception affect the taxpayers' obedience through e-filing utility as much as 0.0283 (0.156 x 0.182)
7. Easiness perception affect the taxpayers' obedience through e-filing utility as much as 0.2819 (0.499 x 0.565)

The path diagram of the above model is as follows:

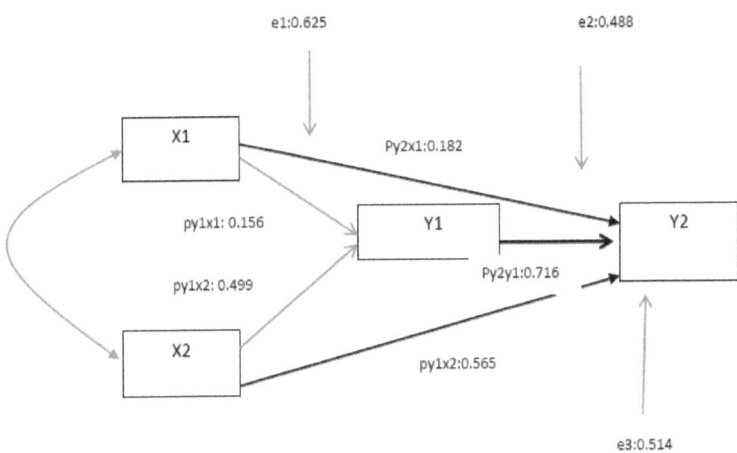

Figure 10.4 A Path Diagram of Relationship among the Variables of Usefulness Perception, Easiness Perception, E-filing Utility and Taxpayers with the Parameter Values

Structural equations for the above model above are:
sub structure 1: $Y_1 = 0.156 X_1 + 0.499 X_2 + e_1$
sub structure 2: $Y_2 = 0.716 X_1 + e_2$
sub structure 3: $Y_2 = 0.182 X_1 + 0.565 X_2 + e_3$

CHAPTER 10

INDIRECT EFFECT CALCULATION

Using an additional software of PROCESS v3.2 for SPSS created by Andrew F Hayes Ph. D we can calculate the indirect in IBM SPSS. Steps to install the program are as follows:

- Download program PROCESS v3.2 for SPSS
- Open IBM SPSS
- Install the program PROCESS v3.2 for SPSS using the "process" file
- The program is ready to use

Example of the Case

In this example we will use the IBM SPSS sample file entitled bankloan.sav. Three variables that we are going to use are:

- Independent variable: Household income in thousand
- Dependent variable: Debt to Income Ratio
- Intervening variable: Credit card debt in thousand

The model of relationship based on theory is as follows:

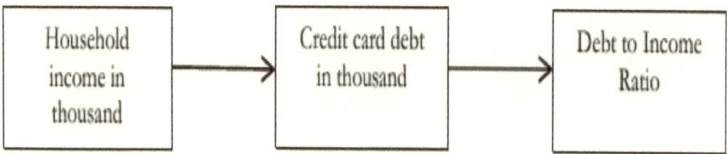

PATH ANALYSIS: DATA ANALYSIS APPLICATION

The path analysis model is as follows

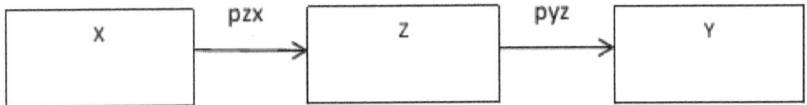

Where:

- X: Household income in thousand (income)
- Z: Credit card debt in thousand (creddebt)
- Y: Debt to Income ratio (debtinc)

Inidrect Effect

Indirect effect in the model is as follows:

IE: X → Z → Y (The effect of X to Y through Z)

To calculate the indirect effect, use the following steps:

- Activate the SPSS
- Open the sample file: Bankloan.sav
- Analyze > Regression
- Select: Process v3.2

The display is as follows

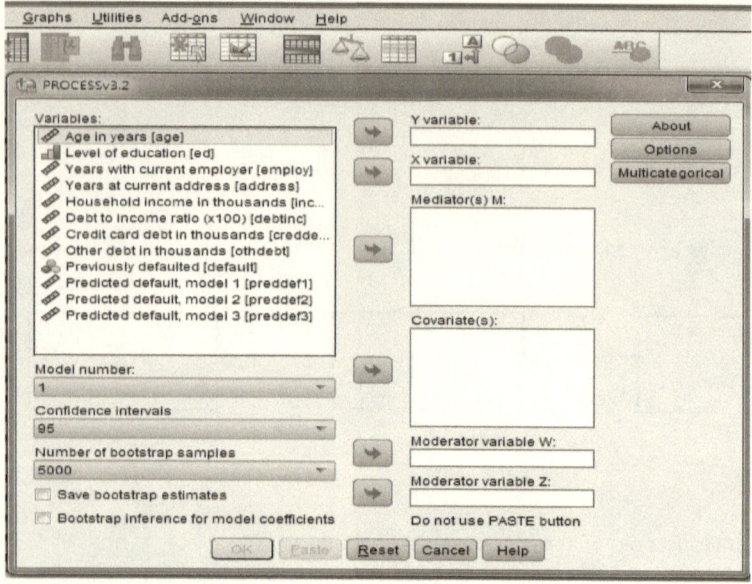

- Move Household income in thousand to X Variable
- Move Credit card debt in thousand to Mediator
- Move Debt to Income ratio to Y Variable

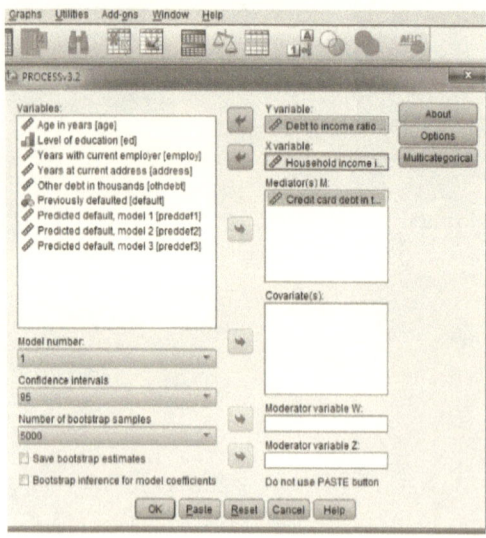

- In Model Number option: select 4

PATH ANALYSIS: DATA ANALYSIS APPLICATION

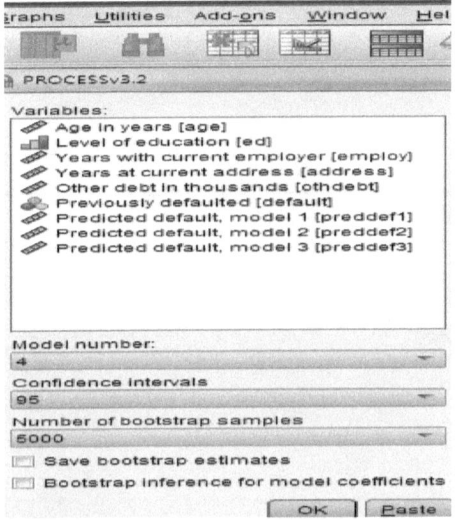

- Select: Option
- Check (v): Generate Code for Visualizing Interactions and Effect SIze
- Conditioning Values: select –SD, Mean, +SD

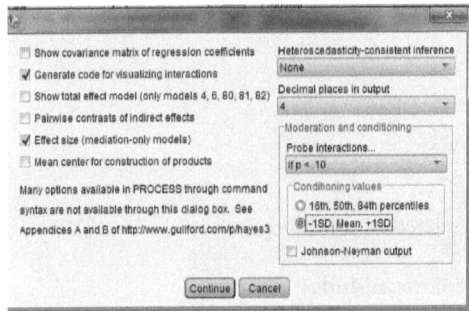

- Continue
- OK

The output is as follows:

```
Run MATRIX procedure:

**************** PROCESS Procedure for SPSS Version 3.2 ***

          Written by Andrew F. Hayes, Ph.D.      www.afhay
    Documentation available in Hayes (2018). www.guilford.c

**************************************************************
Model  : 4
    Y  : debtinc
    X  : income
    M  : creddebt

Sample
Size:  850

**************************************************************
OUTCOME VARIABLE:
 creddebt

Model Summary
       R        R-sq         MSE          F         df1
    ,5515       ,3042      3,1483    370,6870      1,0000      8

Model
              coeff         se           t           p
constant     ,1570        ,0956      1,6420       ,1010       -,
income       ,0304        ,0016     19,2532       ,0000

**************************************************************
```

Main values from the output related to the indirect effect:

Part I: Model 4 (Simple Mediation Model)

- The probability value (p value) for the income is 0,0000 with the t value as much as 19.2532
  The model is correct since the p value as much as 0,0000 < 0,05
- The R square is as much 0,3042 meaning that the effect of income to debtinc through creddebt is as much 0,3042

PATH ANALYSIS: DATA ANALYSIS APPLICATION

```
OUTCOME VARIABLE:
 debtinc

Model Summary
       R        R-sq       MSE         F         df1        df2
    ,6419      ,4120    26,6123    296,7134    2,0000    847,0000

Model
             coeff        se          t          p        LLCI
constant    10,0800     ,2784     36,2030      ,0000     9,5335    1
income       -,0801     ,0055    -14,5410      ,0000     -,0909
creddebt     2,4284     ,0998     24,3229      ,0000     2,2324

***************** DIRECT AND INDIRECT EFFECTS OF X ON Y ***********

Direct effect of X on Y
     Effect       se          t          p        LLCI       ULCI
     -,0801     ,0055    -14,5410      ,0000     -,0909     -,0693

Indirect effect(s) of X on Y:
            Effect      BootSE    BootLLCI   BootULCI
creddebt    ,0739       ,0077      ,0604      ,0910

Partially standardized indirect effect(s) of X on Y:
            Effect      BootSE    BootLLCI   BootULCI
creddebt    ,0110       ,0011      ,0091      ,0135

Completely standardized indirect effect(s) of X on Y:
            Effect      BootSE    BootLLCI   BootULCI
creddebt    ,4237       ,0499      ,3359      ,5300
```

Part II: Outcome Variable (The dependent variable) of Debtinc

- Model Summary: The value of probability (p) is 0,0000 < 0,05 so the the model under study is correct
- Income independent variable: The value of probability (p) is 0,0000 < 0,05, so the income affect the debtincsignificantly through the debtinc. The effect is as much as -,0801
- Creddebt intervening variable: The value of probability (p) is 0,0000 < 0,05, so the creddebt affect the debtincsignificantly. The effect is as much as 2,4284

Part III: Hypothesis Testing

The hypothesis for the indirect effect will as follows:

- H0: Household income in thousand does not affect Debt to Income ratio through Credit card debt in thousand significantly
- H1: Household income in thousand affects Debt to Income ratio through Credit card debt in thousand significantly

Criteria of Hypothesis Testing

- Accept H0, when the probability value > 0,05
- Reject H0, when the probability value < 0,05

Decision

The probability of income to debtinc variable is as much as 0,0000 < 0,05; accordingly reject H0 and accept H1. It means that Household income in thousand affects Debt to Income ratio through Credit card debt in thousand significantly

```
*********************** ANALYSIS NOTES AND ERRORS *****************
Level of confidence for all confidence intervals in output:
   95,0000
Number of bootstrap samples for percentile bootstrap confidence inte
   5000
------ END MATRIX -----
```

The confidence level is as much as 95%. The conclusion is that the Household income in thousand affects Debt to Income ratio through Credit card debt in thousand significantly with the confidence level is as much as 95%.

References

Denis, Daniel J. and Joanna Legerski. (2006). *Causal Modeling and the Origins of Path Analysis.* University of Montana

Duncan, O. D., & Hodge, R. W. (1963). Education and occupational mobility: A regression analysis. *The American Journal of Sociology,* 68, 629-644.

Garson, David (2011) Path Analysis. http://faculty.chass.ncsu.edu/garson

Hair, Joseph F. et al. (2010). *Multivariate Data Analysis: A Global Perspective.* New Jersey: Pearson Prentice Hall

Hayes, F.A. PROCESS procedure for SPSS version 3.2. www.afhayes.com

Johnson, Richard A. and Wichern, Dean W.(2002). *Applied Multivariate Statistical Analysis.* New Jersey: Prentice Hall

Lleras , Christy (2011) *Path Analysis.* Pennsylvania State University Park Pennsylvania USA

Natalia, K(2017). *The effect of Usefulness and Easiness Perception of E-filing and its Impact on Taxpayers' Obedience in Central Jakarta.* Thesis. Jakarta:Tarumanagara University

Olabutiyi, Moses. E.(2006). *A User's Guide to Path Analysis..* Maryland: University Press of America

Sarwono, Jonathan.(2010). *Analisis Jalur Untuk Riset Bisnis* dengan SPSS. Edisi 5. Yogyakarta: Penerbit Andi.

Sarwono, Jonathan.(2012) . *Path Analysis: Teori, Aplikasi, Prosedur Analisis untuk Riset Skripsi, Tesis dan Disertasi.* Jakarta: Elexmedia Komputindo

Streiner, David L. *Finding Our Way: An Introduction to Path Analysis.* Can J Psychiatry, Vol 50, No.2 February 2005

Schumacker, Randall E. and Richard G. Lomax .(1996) *A beginner's guide to structural equation modeling* . New Jersey: Lawrence Erlbaum Associates Inc

Wright, Sewal. (1920). The relative importance of heredity and environment in determining the piebald pattern of guinea-pigs. *Proceedings of the National Academy of Sciences, 6* , 320-332.

Wolfle, L. M. (2003). The introduction of path analysis to the social sciences, and some emergent themes: An annotated bibliography. *Structural Equation Modeling,10*, 1- 34.

Wuensch, Karl L. (2008). *An Introduction to Path Analysis*

ABOUT THE AUTHOR

Jonathan Sarwono currently is the Director of Quality Assurance in International Women University Bandung, Indonesia. He is also a lecturer in some universities in Bandung and Jakarta as well as a trainer on statistics in several companies in Jakarta. So far 50 books have been written about statistics using IBM SPSS, EVIEWS, LISREL, SmartPLS, AMOS and STATA. Beside that, he also writes several books on Research Methodology and Information Technology. The books have published both in the country and overseas as well as sold internationally. He can be contacted through his web site, **http://www.jonathansarwono.info** or email, jsarwono007@gmail.com.

www.ingramcontent.com/pod-product-compliance
Lightning Source LLC
Chambersburg PA
CBHW030633220526
45463CB00004B/1507